# A History of Geology

## Gabriel Gohau

revised and translated by
Albert V. Carozzi and Marguerite Carozzi

Rutgers University Press
New Brunswick and London

This work was published with the aid of the French Ministry of Culture and Commu-
nication. The publisher, translators, and author gratefully acknowledge the Ministry's
assistance.

Library of Congress Cataloging-in-Publication Data

Gohau, Gabriel.
    [Histoire de la géologie. English]
    A history of geology / by Gabriel Gohau ; revised and translated from the French
  by Albert V. Carozzi and Marguerite Carozzi.
      p.  cm.
    Translation of: Histoire de la géologie.
    Includes bibliographical references and index.
    ISBN 0–8135–1665–X (cloth)—ISBN 0–8135–1666–8 (pbk.)
    1. Geology—History.  I. Title.
QE11.G64   1991                                   90-47755
550′.9—dc20                                       CIP

British Cataloging-in-Publication information available.

# Contents

Contents

Contents

# List of Figures

# Translators' Foreword

Translating this book into English was a very rewarding experience for both the author and the translators, who worked together closely, page by page, on both sides of the Atlantic.

The text presented here is not merely a translation of the original French; it is a revised edition of Gabriel Gohau's work in English. Besides a new general emphasis toward the American reader, a more scholarly approach has been achieved by providing complete bibliographical references, by sharpening discussions and concepts, by updating topics, and by adding new illustrations and an extensive glossary (terms included in the glossary are boldfaced the first time they are mentioned).

The author and the translators hope these improvements will provide readers with a better understanding of the history of geology—a goal to which the three of us are deeply committed.

Albert and Marguerite Carozzi
University of Illinois at Urbana–Champaign

# A History of Geology

# Introduction

A scientist, however little inclined toward philosophical speculations, who wishes to write the history of his or her field cannot
ignore the question, When did this science start? Geology became a
self-sufficient scientific discipline with its own research and teaching institutions only at the end of the eighteenth century. However,
ever since prehistoric times people have extracted both raw materials
and tools from the earth and have invented mythical stories about
the creation of the world. The goal of geology is to explore that part
of the earth that is accessible to direct observation and to establish its
history. Hence, human beings have always done some kind of "geology"—even when the word itself did not exist.

The fashion of extending the term *science* to all past investigations that have covered the same subject is not without serious draw-
backs. For instance, it gives to mythical explanations of the origin of
the world an unwarranted scientific status. We shall therefore limit
our discussion of ancient "geology" to the answers given by the great
philosophers of ancient Greece.

Their early speculations referred, however, to the cosmos as a
whole and emphasized cosmology rather than geology. Only after the
astronomical revolution of Copernicus and Galileo did thoughts
about the earth become dissociated from those about the universe. It
is thus proper to start the history of geology at the time of this
dissociation.

In 1681, Thomas Burnet (1635–1715) used the phrase "theory of the Earth" to name his work on the formation and functioning of the earth. He had been influenced by a schematic account sketched a quarter of a century earlier by René Descartes (1596–1650) in the fourth part of his *Principes de la philosophie*. Although this history could have begun with Descartes, I have included, by way of prologue, two chapters about ancient debates on continuity versus discontinuity. These debates, which set the Stoics against the disciples of Aristotle, set the stage for more modern discussions.

The sixteenth- and seventeenth-century theories of the earth represent the first act of our history of geology. However, they only explained the formation of the earth rather than its true history. Furthermore, if we take collective work to be the characteristic trait of science, these theories belong, at best, at the threshold of this history because the phrase "theory of the earth" really only describes an individual work—each author's own theory.

At the end of the eighteenth century, the scenery changed suddenly and Act 2 began. A new term, *geology*, described an enterprise common to all who were interested in the earth's past. Geology was not the only science to emerge as a distinct field just then. The term *biology* was first used around 1800, when it was recognized that the division of natural history into three kingdoms (mineral, plant, and animal) had to be replaced by a division between the organic world and the inorganic world. This was also a decisive time for chemistry, thanks to Antoine-Laurent Lavoisier (1743–1794) and for study of electricity, thanks to Charles Augustin de Coulomb (1736–1806). Modern science was perhaps born twice: first, in the generations that joined Galileo to Newton, and a second time as industrial society emerged.

At this turning point the term *geology* appeared in the works of two Genevan naturalists. In 1778, Jean-André Deluc (1727–1817) stated that the term *geology* would be more appropriate than *cosmology* to describe "knowledge of the Earth" (*Lettres physiques et morales sur les montagnes*). In 1779, the word *géologie* also appeared in the introduction to the first volume of the *Voyages dans les Alpes* by Horace-Bénédict de Saussure (1740–1799). He became famous among the general public for promoting the first attempts to climb Mont Blanc and for reaching the summit himself on August 3, 1787.

The two Genevans had not coined the term on their own, although they were the first to use it professionally. The Latin word *geologia* had been used in 1473, but without the modern meaning. In 1735, Benjamin Martin mentioned geology in *The Philosophical Grammar* as a field that also included the study of plants and animals. Denis Diderot (1713–1784) introduced it in his *Encyclopédie* in 1751 along with other new terms.[1] Soon *geology* not only entered encyclopedias but was also taught in many schools.

The National Convention—a French revolutionary government, 1792 to 1795, led by Robespierre among others—built a new institution in 1793 to replace the former Jardin du Roi (the king's garden founded at the time of Louis XIII). This National Museum of Natural History was awarded twelve teaching positions, including one for geology. The geology position was given to the naturalist B. Faujas de Saint-Fond (1741–1819). Mineralogy had already been taught at the Jardin du Roi by L.J.-M. Daubenton (1716–1800), friend and associate of the great naturalist Buffon, and continued to be taught in the National Museum of Natural History, so two teaching positions now covered earth sciences. More important, the term *geology* was used to designate one of them.

Geology was introduced less rapidly in other institutions. Napoleon founded the University of France in 1808. Four faculties (fine arts, law, medicine, theology) were established at Paris. The faculty of fine arts, divided into letters and sciences, was organized in 1809. Nine positions were awarded to the faculty of mathematical and physical sciences. One of them was given to René Just Haüy (1743–1822), the founder of modern mineralogy. Haüy was assisted by Alexandre Brongniart (1770–1847), who was officially in charge of geology but in fact taught mainly paleontology, his own particular interest.

Brongniart was, as we shall see, a learned man of great insight who can be regarded as one of the founders of **biostratigraphy.** However, the teaching of geology remained subordinate to mineralogy and became independent only in 1831. The position was then given to Constant Prévost (1787–1856), who had founded the year before, with Ami Boué (1794–1881), the Geological Society of France.

The following year, Georges Cuvier (1769–1832) died suddenly. He had held a chair of natural history at the Collège de France, where he had taught mainly biology. He was assisted by J.-C. de Lamétherie (1743–1817), who dedicated himself to geology. The geologist Léonce Élie de Beaumont (1798–1894) was entrusted with Cuvier's position.

Five years later the position was divided into two, leaving to Élie de Beaumont the teaching of the natural history of inorganic beings and creating a chair of natural history of organic beings for G. L. Duvernoy, a student of Cuvier.

This review (a little tedious, but the reader will encounter all of these names again) is simply intended to show how the teaching of geology developed in one country, France, during the first fifty years of the science. By 1832, geology was taught at the National Museum of Natural History, at the Faculty of Sciences of the University of France at Paris, and at the Collège de France. The School of Mines was opened in 1783 and restructured in 1794 after a brief closure. French geologists formed a professional society in 1830 as their colleagues in England had done in 1807.

The second act of this history of geology, which began with much ado around 1800, continued all through the century and culminated in the great synthesis developed by Eduard Suess (1831–1914) around 1900.

The scenery then changed quietly. The actors themselves were not immediately aware of the change. Fifty years, from 1912 to 1962, separate Alfred Wegener and Harry H. Hess, who together prepared a new revolution. The stage was set for Act 3 when the ocean took the place of the continent at center stage. This act is still being played out today, and we are still in its opening scenes. I have written it here as an epilogue since we cannot know what decisive scenes will be played by our descendants.

One more point. Although I have contrasted mineralogy with geology, the history worked out in this book is that of historical geology alone. Indeed, historians of geology may either study the history of mining, metallurgy, crystallography, petrography—in short, all the subdisciplines dealing with rocks and minerals—or they may be interested in the history of observations of geologic features, in the development of theories that explain the origin of these features, and in the history of the observers themselves and the institutions to which they belonged.

In this history of geology, I have chosen the second course, to emphasize philosophical debates among theoreticians rather than the work of practitioners. It is true that the role of practitioners is far from insignificant in the development of a science. We shall encounter them here and there, especially in connection with the industrial revolution. However, I have chosen philosophical debates because they better reveal the historical character of geology itself. It is a historical or "palaetiological" science, as William Whewell, a geologist and philosopher of science, called it to stress its originality. Considering historical sciences such as geology, he said, "The investigations which I now wish to group together, deal not only with the possible, but with the actual past."[2] In other words, in comparison to astronomy, for instance, which studies the universe on the basis of computerized mathematical calculations, geology studies real objects—rocks, fossils, folded layers, and so forth—to understand the past of the earth.

Science has its regular patterns, which are translated into laws; whereas history has its events, which are found in its archives. Geology exists at the zone of interaction between science and history. The interplay between the two approaches—scientific and historical—raises fascinating philosophical questions for geology, and this, to some extent, determined my emphasis on historical geology here.

# Chapter 1

# The Beginnings

## Introducing Geology

Even though modern geology emerged during the last years of the eighteenth century, the first chapters of this book treat a kind of pre-history of the earth sciences. The concepts I discuss in these chapters are therefore not those of our own time, and the reader as well as the author will be tempted to indulge in anachronisms. Yet it would be absurd to lay aside our present knowledge, both because it is impossible to do and because ideas in the history of every science tend to recur in some form.

If we tried to reconstruct the ideas of the ancients without introducing any modern concepts, we would (at best) do ancient science, but not history. Each glance into the past is necessarily retrospective. However, we must not let our own science dominate us. Gaston Bachelard, French philosopher and historian of science, has said that the fault of empiricists is that they consider every glimmer of light as true knowledge.[1] We should not behave like empiricists, searching in the past for the seeds of modern science. We dare not consider our predecessors simply as the precursors of our contemporaries. Indeed, every civilization in the past had its own rationality and objectives, with no thought of laying the way for our present knowledge. We must not overly finalize the path of science. The goal of a historian should be to find fundamental trends in retrospect.

Of course, in order to write the history of a science, it is imperative to have some knowledge of that field. Let me therefore give the reader a short introduction to geology.

By tradition, geology has been divided into various disciplines according to subject matter. Petrography describes and classifies **igneous**, **metamorphic**, and **sedimentary rocks**. Petrology deals with the origin, occurrence, structure, and history of rocks. Volcanology studies volcanic events, whereas sedimentology studies sedimentary environments and the processes of deposition of sedimentary rocks. Paleontology reconstructs fossils. Stratigraphy reads in the layers (strata) of the earth the past history of the globe. Stratigraphers use fossils to date rock units; hence, stratigraphy relies on **biostratigraphy**. Stratigraphy thus reconstructs past successive geographic landscapes that lead to paleogeography. Finally, structural geology studies deformation of strata generated by movements of the earth's crust. It recognizes the uplifting of ancient mountain chains, or **orogenies**.

However, geology can also be divided in other ways: for instance, into physical geology and historical geology. Physical geology treats geological phenomena of the present, whether external (erosion, **sedimentation**) or internal (volcanoes, seismic activities). Historical geology reconstructs the earth's past. This division emphasizes that geology is different from other sciences in its historical character. (Cosmology and evolutionary biology have this characteristic too, but to a lesser degree: when evolutionary biology reconstructs the past, it becomes paleontology.) The most important distinctive feature of historical geology—and one that I shall stress—is the geologist's necessity of assuming that the earth in the past submitted to the same natural laws as it does in the present: the principle of **uniformitarianism.**

Without this axiom, it would be impossible to reconstruct the past because the evidence (equivalent to the archives of historians) of former eras can be understood only in comparison to phenomena we observe today. Applying this principle to time is, in fact, completely analogous to physicists' assumption that physical laws discovered in our microcosm are universally true throughout space. Thus stated, the principle of uniformitarianism would appear to require no further explanation. We shall see, however, that accepting the principle was not so simple. After all, why should geological phenomena have always occurred with the same intensity? Why should every kind of past event still occur today? All through these pages, we will encounter the difficulties that geologists have had with this principle.

Early concepts of geological cycles illustrate this dilemma. A geological cycle, as defined today, consists of an endlessly repeated succession of three major processes: uplifting, erosion, and sedimentation. Although these processes were not specifically characterized before James Hutton (chapter 9), the ancients were aware of at least two: erosion and sedimentation. Nevertheless, many ancient natural philosophers also believed in catastrophic events capable of drastically changing the surface of the earth, a belief rather removed from true uniformitarianism. The problem of seashells found far from the shore offers a glimpse of their thinking on this subject.

## Shells Found Far from the Shore

One of the difficulties in the teaching of geology lies in our inability to conceptualize the immense duration of the earth's past. Changes at the surface of the earth are measured in millions of years, and the complete history of our globe amounts to billions of years. We feel just as much at a loss with these enormous chronologies as with the immensity of space in astronomy.

Nevertheless, nature provides visible proofs of changes that have occurred at the surface of the earth. The ground on which we walk is very often an ancient seafloor. However, it has sometimes undergone such great changes that we can no longer recognize its origin. Such is the case for areas with igneous and metamorphic rocks. Their layers have lost their original appearance after having been buried in the depths of the earth's crust and undergone **metamorphism** by the action of heat and pressure.

Other rocks, however, have preserved visible traces of their formation in water. Everyone has seen fossils, the remains of living beings in stone. They occur inside parallel layers, which were evidently former deposits of **sediments** later **indurated** and turned into rocks. Today, finding shells of marine animals far from present shores catches the imagination of the collector-paleontologist. Such finds are, to the general public, the most convincing evidence of past displacements of shorelines.

Ancient writers took notice of fossils, of course. In the fifth century B.C., Xanthus of Lydia saw shell-shaped stones far from the sea and concluded that areas where they occurred had once been submerged beneath the sea.[2] These observations also led Herodotus, the famous geographer (ca. 484–420 B.C.), to consider Egypt a former marine bay. The Roman poet Ovid (43 B.C.–A.D. 17) recalled the Pythagorean line: "Often sea-shells lie far from the beach, and men have found old anchors on mountain-tops."[3]

These writers were unaware of the age of the fossils they had found. They could not guess that these former beaches were many millions of years old. And the anchors, however impressively ancient they looked, could not have come from ships wrecked in ancient seas because, in the fossils' distant past, neither ships nor human beings existed. But it is too easy for us to be ironic. In the following centuries, opinions on the origin of shells were not always as clearheaded; these classical authors deserve praise for their perceptiveness.

## Strabo versus Eratosthenes

But how was it possible to explain these ancient seashores? Let us follow the debate between two illustrious geographers. One is Eratosthenes, a learned man of the third century B.C., contemporary of Archimedes, and best known for being the first to measure the circumference of the earth. Eratosthenes was not only an astronomer and geographer but also librarian at the famous library of Alexandria (founded in 332 B.C. by Alexander the Great).

The other author, Strabo, was born a few centuries later in Cappadocia, in the middle of the first century B.C., roughly a contemporary of Julius Caesar. His multivolume *Geography* describes the world as known to the Romans.

Eratosthenes believed that the sea level of the Mediterranean had lowered during the opening of the strait known by the ancients as Pillars of Hercules (Strait of Gibraltar); that this strait had previously been closed as had the Pont-Euxin (Bosphorus Strait), which connected the Black Sea to the Mediterranean Sea; that the two seas had been higher than at present, forming two basins above the level of the ocean; that when the straits opened the basins were partially emptied; and that proofs of the former levels were found in the occurrence of fossils and of "wreckage from seafaring ships" above the present sea level.[4]

Strabo disagreed with Eratosthenes. He did not believe that the water level of the closed seas could ever have been at higher levels than in his own time. And to such explanations, which he called "chimera," he opposed "real causes," for which he searched in natural events. According to Strabo, "earthquakes, volcanic eruptions, uplifting of submarine floors, on the one hand, and sudden collapses or landslides, on the other," raise sea level in the first case and lower it in the second.[5]

It is not difficult for a Mediterranean inhabitant to visualize volcanic eruptions and volcanoes causing **tsunamis** of more or less catastrophic nature. For instance, several authors believed that the splendid Minoan civilization on the Island of Crete may have been wiped out by such a catastrophe—an event sometimes identified with the eruption that destroyed the ancient volcano of Santorin in 1500 B.C. Such events baffled people and nourished legends like the one about Atlantis.[6]

However, Strabo did not believe in legends. He used an observable phenomenon (tsunami) to explain another established fact (the presence of fossils far from the shores). The only objection we might have today is that modern seismic activities cannot explain the deposition of shells, which implies, as Strabo himself acknowledged, that the sea "once covered for periods of variable length" part of the continents. We find here the first distortion of the principle of present-day causes: the ancients liked to invoke phenomena of a greater intensity than those known today. Eratosthenes' opinion is not satisfactory either, since he does not explain what catastrophe opened the straits— but we should remember that we only know his thesis through Strabo, who was perhaps not too eager to explain Eratosthenes' ideas fully.

## Degradation of Mountains and Time

The debate between Eratosthenes and Strabo brings us to another set of concepts: the opposition of reversible variations and irreversible changes. The openings of straits lead directly to the irreversible movement of the sea because the sea level of intracontinental basins can only become lower (an irreversible change). However, movements of the seafloor can cause the sea level to rise and fall (a reversible variation). These movements may in turn cause advances and retreats of shorelines.

This opposition between reversible variations and irreversible changes is at the heart of the discussions among philosophers of ancient Greece. Erosion that causes rocks to fall, to break up, and to be carried away as debris leads to changes at the surface of the earth

---

# Aristotle and His Adversaries

Aristotle, born in 384 B.C. at Stagira (this is why the famous philosopher is often called the Stagirite), was tutor to the future Alexander the Great. He died in 322 B.C., one year after Alexander, at a time when the empire of the conqueror was split up. Aristotle was also a contemporary of the Athenian statesman and orator Demosthenes.

Aristotle studied first in Plato's Academy in Athens and later founded a rival school, which he named Lyceum. He lectured chiefly while walking in the Lyceum, hence, his students were sometimes called Peripatetics (from the Greek "to walk around").

Athens was then at the peak of its reputation. Not much later, however, the city lost its supremacy: this marked the end of the Hellenic period. Alexander's conquests led to the creation of rival cultural centers, in particular the illustrious one of Alexandria, which superseded Athens in the third century B.C. This was the beginning of the Hellenistic period.

Nevertheless, Athens remained for Euclid and his contemporaries an important center of philosophical inquiry. The Stoic school settled there after the death of Aristotle while Theophrastus was in charge of the Lyceum. Zeno of Citium (ca. 336–264 B.C.), the founder of the Stoics, was born half a century after the Stagirite. His most important disciples were Cleanthes and Chrysippus.

Epicurus (341–270 B.C.), only five years older than Zeno, founded a third school in Athens. The physical basis of his system rested on the existence of particles called *atoms*, a name borrowed from an ancient

---

that cannot be ignored by the inhabitants along the Mediterranean shores.

The observation that served as the starting point for debates on the geological cycle among the ancients is one that anyone can make. I have already stressed that finding fossils inside rocks was quite common. An even more frequent, even daily event was the observation of the devastating effects of erosion. To find fossils, you must look in a region with sedimentary rocks. Fossil hunting would be a waste of time in an area of outcropping igneous and metamorphic rocks, that is, rocks that had undergone the action of subterranean forces. However, all landscapes show the effects of degradation of continents. Of course, these effects are more visible in mountains because of their size. Nevertheless, even the quietest river in the plains sometimes carries tons of mud, proving that landslides or collapses of mountains occurred upstream. The least attentive observer notices that

*(continued)*

philosopher, Democritus (fifth century B.C.). Only a few excerpts from Epicurus's work have come down to us, mostly through the account of his illustrious disciple, the Latin poet Lucretius (ca. 99–55 B.C.), in his famous *De rerum natura* (Of the nature of things). I should note that the surviving writings of the Greek authors I mentioned earlier are but a small portion of those they are said to have written. Although numerous works by Aristotle are known today, they represent only a small fraction of the hundreds of works attributed to him.

Our knowledge of the ideas of Epicurus and the Stoics comes from the works of commentators, often of secondary rank, called *doxographers* (from the Greek *doxa*, opinion, and *grapher*, one who writes), who collected and discussed the opinions of notable philosophers. The first among them is Theophrastus, Aristotle's successor and author of *Physicorum opiniones* (Opinions of philosophers on nature) in *Doxographi Graeci*. Theophrastus was certainly the source used by other doxographers, authors less illustrious than himself, whose treatises sometimes gave a distorted view of the original doctrine.

Turning to later disciples, whose work is better known, could overcome this difficulty, as long as these distant disciples are more faithful to the original than the commentators. I have chosen to present Epicurus through the poem by Lucretius. I might also have explained the Stoics through the biased view of their Roman disciples, Seneca, Epictetus, and Marcus Aurelius, in particular Seneca (the two others being essentially moralists), who in *Naturales quaestiones* (Natural questions) studied terrestrial phenomena.

these muds and silts flow down the valleys toward final deposition in the sea. A more inquisitive mind may ask why continents undergoing degradation do not become **peneplains** and why oceans are not filled with all the debris from the continents. I shall return to these questions shortly.

Geology treated by the ancients obviously rested on everyday observations. The prehistory of all science rests on a common stock of knowledge. Only when true sciences are born do the bases of observations and experiments change and move out of the realm of everyday experience.

Of course, not everyone was capable of understanding the ideas that ancient philosophers drew from these everyday observations. It is not enough to observe the floods of a tumultuous river carrying muds in suspension to arrive at the concept of the geological cycle.

Furthermore, in order to conceptualize that these phenomena

lead to a general peneplanation of the terrestrial globe, it is necessary to imagine the long time span assumed by modern geology. If the earth were only six thousand years old, as seventeenth-century theologians and natural philosophers believed, processes of land erosion and marine deposition would have caused very little change to that earth. If it was going to be destroyed in the very near future by an act of God, as many in the seventeenth century feared, it could not have lost its morphology before the final Apocalypse.

The ancients, however, did not believe in an impending end of the world, nor in its recent creation. Aristotle, for instance, believed the world to be eternal. This is why he was forced to admit that the earth must repair the effects of degradation on its relief. Therefore, as one of his disciples said, "if some mountains collapse . . . others must be formed."

Confronting the school of Aristotle (the Peripatetics), the Stoics upheld that, on the contrary, "the whole whose parts are subjected to destruction, must also face destruction." Therefore, since rocks in mountains (part of the universe) collapse, the entire world (the whole) must also perish. However, the universe is capable of re-generation, of re-creation, so that, in the concept of successive worlds, some sort of eternity does exist. To adopt a metaphor of the life sciences: the eternity of lineage of the Stoics was opposed to the Aristotelian eternity of the individual.

## Catastrophes or the World in Equilibrium

The ideas of the Peripatetic and the Stoic schools represent two opposite viewpoints on the age of the earth. Indeed, for Aristotle the world is eternal, and if it shows any signs of decay, it must necessarily also show signs of rejuvenation in order to endure. Therefore, whereas some places will "dry up and become old," others, "full of life," shall become moist again.[7] Even though the sea evaporates, it will never dry up because waters are constantly being formed. Similarly, the saltiness of the sea remains constant because, while salt evaporates (another erroneous viewpoint), "dry exhalations" bring new salts to the sea.

The Stoics had a completely different view, holding that the earth is doomed to perish. Therefore, signs of decay observed were proofs of an impending end. However, they believed, at the same time, that the earth would regenerate. After a "universal conflagration, which reduces every element into fire," the earth would be born again. "All beings and events are thus returning eternally." And this phrase "eternal return" must be taken here in its most literal sense since, according to Nemesis, disciple of Zeno, "There shall be again a Socrates, a Plato . . . and this restoration shall not only occur once but several times: or rather, all things shall be eternally restored."[8]

Thus, in the view of the Stoics there is no less eternity, if one may put it that way, than in that of the Aristotelians. And this statement could be extended to the third doctrine of Greek antiquity, that of Epicurus and his disciples. The Epicureans also believed that the earth is periodically destroyed and re-created. Their view differed, however, from that of the Stoics by the place held by chance in the combination of the elements. Their belief in chance, or randomness, did not, in fact, allow the recurrence of the same events.

## The Formation of the World

Aristotle and his disciples did not tell us how mountains are constantly being repaired and thus maintain their reliefs. The Stoics, who favored the idea of periodic regeneration of the earth, explained neither repair nor the preceding conflagration. When they mentioned fire as the agent, it was meant in the sense of a purifier because their preoccupations were above all moral ones. Seneca, a later disciple, was more precise when he said that the earth may be destroyed either by the deluge or by fire; however, his ideas were borrowed from a contemporary at the beginning of Christianity.[9]

What is so interesting in Lucretius are his details on the formation of our earth. I emphasize our because the Epicurean earth has some sort of a personality, which that of the Stoics lacks. Formed by a random combination of atoms, this earth is more or less a fortuitous one, which differs from the repetitive universe described by Nemesis. And since every world is somehow unique, the formation of ours becomes of interest. To explain the difference between Lucretius and the Stoics, one could say—referring to the metaphor used above— that the preservation of lineage is no longer of interest, but the development of the individual is crucial.

Lucretius made it quite clear that this earth, together with "the sky and the land and the sea," is nothing "in comparison with the sum total of all the sums," but that he would nonetheless only speak about this earth.[10] Similarly, in the eternity of time, he set off the duration of our earth.

He explained that the earth appeared out of chaos in which atoms formed some sort of "prodigious hurly-burly mass compounded of all kinds of primal germs" when "like with like began to join." "At that time the lands separated from the lofty heavens and the ocean spread over its bed."[11] Thereafter (he spoke only of the surface of the earth), while condensation of matter continued, the ground began to sink. However, this sinking was hindered in some places by the accumulation of rocks; mountains were formed in these locations. Indeed, only in mountains, and not in the plains, can bare rocks be seen.

Lucretius influenced philosophers and naturalists of the seventeenth and eighteenth century who tried, as he did, to explain the *formation* of the earth. The model proposed by Descartes, the first of these philosophers, is quite different from Lucretius'. Descartes himself strongly refuted the atomists' concept of a vacuum. Nevertheless, the efforts of both philosophers were directed along similar lines, and their models have, as we shall see, various points in common that cannot be disregarded.

If, in this first chapter, I have mentioned developments of ideas that may be judged rather farfetched by the reader, I have done so because they can be considered prototypes of more modern theories. Lucretius had followers among naturalists during the seventeenth and eighteenth century. The debate between the Stoics and Aristotle resembles the controversies between catastrophists and uniformitarians at the beginning of the nineteenth century. The Stoic view of an earth that evolved in cycles, with phases of destruction followed by phases of restoration, must be seen against Aristotle's earth, which had so many overlapping cycles compensating for one another that the earth as a whole remained in equilibrium. The same conflicts arose between supporters of the catastrophic concept of orogenic phases separating periods of rest, and those who favored the uniformitarian view of more or less continuous movements. The term *steady state*, used by British and American historians to describe the uniformitarian model of Charles Lyell (1797–1875), emphasizes the idea of equilibrium, which is also characteristic of the Aristotelian concept.

## Physical World and Moral World

Everyone has heard the legend of the sinking of Atlantis, a legend from which believers in science fiction and folklore still try to extract evidence of the existence of lost civilizations. It is possible that the reader has already wondered why Plato—who first related that legend—has not yet been mentioned in this review of ancient doctrines. His relatively moderate catastrophism comes close to that of the Stoics, or, more precisely, falls between the Stoics and Aristotelian uniformitarianism.

Plato's catastrophes are moderate in the sense that, even though they destroy all humanity and "everything on earth," they are nevertheless limited to the surface of the earth. Thus, the whole world is not destroyed periodically, only living creatures. Moreover, Plato's agents—earthquakes and floods—are phenomena belonging to present-day causes, which are exceptional only by their intensity. In other words, the world endures these catastrophes as did Aristotle's earth. The essential difference between Plato and Aristotle is Plato's belief that the world was created and is thus not eternal.

It may seem odd that Plato, the earliest of the ancient philosophers analyzed in this chapter, comes last, though by chronological order he should have been mentioned first. I discuss Plato last because in *Timeaus* and *Critias* he is not really describing the earth, but its inhabitants. Plato's intent was moral: for him, catastrophes were divine designs to purify the earth.[12] When God ruled the earth, it was the golden age; when God abandoned it, destruction set in. Catastrophes purged humanity of all the disorders contracted while God was absent.

Plato's principal preoccupation was thus clearly a moral one. But this raises a very important issue for us: how can historians of the earth sciences claim to study natural history in doctrines that are essentially moral and in which natural phenomena are only a frame for speculations on human activity and on the development of society? To what extent can this frame be separated from the purpose it served?

This moral framework certainly represents an obstacle to the investigation of Plato's ideas on natural phenomena. As modern science was born, scientists tried with increasing success to get rid of ghosts, of dreams about the physical world that blurred their objective view of natural phenomena. However, for this early period, long before the scientific revolution (or revolutions), ambiguities prevail and historians of a science cannot easily say what should count as their science.

At any rate, historians must not judge this distant science by the standards of modern science. To do so would run the risk of distributing laurels to doctrines that resemble our modern ideas only by chance, or for reasons that have nothing to do with scientific observation.

Compared to the later ideas of Aristotle and Zeno, Plato's account of ancient catastrophes seems balanced and harmonious. Yet, unlike them, he could not wholly dispense with mythologies and interventionist gods. His world was not destroyed by catastrophes because the same divine will that had created it was also preserving it. Aristotle and Zeno produced more rational systems. To us, atomism looks like the most modern concept because it is free from primitive animism.

Nevertheless, the first Christians felt closer to Plato than to the other writers. The fate of the ancient doctrines during the Middle Ages, both in the Christian and the Islamic world, shall be examined in the next chapter.

# Chapter 2

# At the Center
# of the World

## The Historical Framework

Alexandria, the capital of Egypt, was the main cultural center during the first centuries of the Christian era. Galen, famous physician of the second century A.D., was born at Pergamum, Asia Minor (intellectually the rival city of Alexandria), but he completed his studies in the Egyptian capital. His contemporary, Ptolemy, the geographer and astronomer whose system lasted until Copernicus thirteen centuries later, lived in the same city.

The Roman Empire started to disintegrate during the fourth century and foundered in the following century under the pressure of the great migrations from the East. Before the fall of the Roman Empire (A.D. 476), the western part of the empire became Christian. Constantine, emperor from A.D. 306 to 337, converted to the new religion. He convened the Council of Nicaea (A.D. 325), where the foundation of the Christian doctrine of the Trinity was laid.

This religious evolution is of interest to historians of geology. Indeed, as mentioned above, Aristotle believed in an eternal world. The Church could not accept such a philosophy, yet rapidly became accustomed to Plato's creed. Origen (A.D. 185?–?254), one of the fathers of the Church, was able to combine Plato's concepts with Christian dogma. Saint Augustine, bishop of Hippo in North Africa (A.D. 354–430), told in his *Confessions* how he was led to

Christianity through Platonism—or at least as Platonism was understood by its distant disciples. Plotinus (A.D. 205?–270), who was
sometimes confused with Plato because of the resemblance in name,
was one of these disciples.

Plato's God, endowed with a personality and free will, was relatively close to the Christian God, much closer than the totally abstract "motionless mover" of Aristotle. The immanent God of the
Stoics also became popular with practitioners of the new religion,
and the Stoic philosophy influenced Christian cosmology. All that
Christians needed to do was to replace the Stoics' notion of successive worlds with a unique act of creation and destruction by the
Christian God.

Nevertheless, the astronomical doctrines of the ancients were not
readily adopted because no one who accepted the Bible could literally believe that the universe was made of concentric spheres around
the terrestrial globe. Indeed, the Old Testament image of the world
envisaged a flat earth floating on the ocean; above it, the firmament,
held up by pillars; and, in turn, the waters above supported by the
firmament. It is true that, at the beginning of the fourth century, Lactantius, tutor to the son of the Emperor Constantine, tried to ridicule
the concept of a spherical earth. However, the Church never officially
accepted his doctrine. As Thomas Kuhn put it, "There was no Christian unanimity about cosmology during the first half of the Middle
Ages."[1]

## The "Luminaries" of the Middle Ages

In reality, Church leaders tried to turn attention away from science
and secular, pagan knowledge. To pay so much attention to "physics" was useless, if not downright dangerous, to the soul's salvation.
For the purposes of salvation, knowledge of the Scriptures was quite
sufficient. Church leaders in fact praised "beneficial" ignorance.

This position of the Church may come as a surprise to us since it
seems to contradict two clichés well anchored in modern minds.
First, when we remember how bitterly the Church fought the idea of
the Copernican universe, forcing upon Galileo his pathetic recanta-
tion of 1634 (chapter 3), we might conclude that the Church had always supported Aristotle's world of concentric spheres. That was not
the case. A few faithful readers of Scripture like Lactantius simply
felt uncomfortable with such an image of the world; more thoughtful

theologians argued that such an arrangement of the cosmos seemed to put constraints on the omnipotence of God. In the thirteenth century, the argument about God's omnipotence came up again when the question was raised whether our earth was the only one or whether other worlds existed elsewhere. The bishop of Paris maintained, even against Thomas Aquinas himself, that it would limit divine power to maintain that God could not have, at least in principle, created several worlds.[2]

According to the second cliché, the ignorant folk of the Middle Ages believed (before the travels of the great navigators at the end of the fifteenth century) that the earth was flat. In fact, most medieval natural philosophers never denied the existence of the antipodes. Of course, those who discussed such matters comprised a very limited circle of scholars, but it is the only one we are considering here. If we had to take into consideration popular beliefs, it would be quite difficult to say when the idea of a spherical earth entered the realm of common knowledge. Indeed, in light of a recent poll by two French sociologists, which revealed that one third of the French people in the 1980s still believed that the sun revolves around the earth, we can only be baffled about the inconceivably slow diffusion of scientific knowledge.

Nevertheless, we should not believe that the Christian Middle Ages were a time of enlightenment. Although the name "the Dark Ages" is not entirely warranted, it was, at least in the beginning, a time when knowledge regressed precisely because of the Church's disdain for science.

It was the Islamic world that gathered the inheritance of Greek civilization. By A.D. 1000, Arab and Byzantine scholars had already translated the major scientific works of the ancient world, treatises that only became known in the Christian West in the twelfth century (and some not until the Renaissance). Aristotle's work was analyzed by philosophers such as Avicenna, born a little over a thousand years ago (A.D. 980), and Averroës, who lived a century later, whose commentaries in turn influenced medieval Christian natural philosophy.

Starting in the twelfth century, the works of commentators and of Greek authors reached the West. They were translated from the Arabic into Latin. Later on, translations were made directly from the Greek. In the thirteenth century, Aristotle's ideas slowly entered the Christian worldview. Yet another cliché should be abandoned here: If modern science, with Galileo and Descartes, had to free itself from

---

# Flying Fossils

People in China noticed fossils very early, mentioning their existence in a text dated A.D. 527. During the Sung dynasty (tenth to thirteenth century), a period of high cultural achievement, it was known that mountains were formed under water.

Among the Chinese legends at the beginning of the Middle Ages is one about the spirifer, a shell with a transverse elongated shape that vaguely recalls short wings. Around the year 375, an author mentioned a mountain with "stony swallows" that flew during thunderstorms. An author of the twelfth century, who obviously no longer believed the story, observed (after having marked the fossils) that rain made them fall. This might have kept alive the legend that the fossils were flying.*

---

*René Taton, ed. *Histoire générale des sciences*, 4 vols. in 3 (Paris: Presses universitaires de France, 1957–1964), 1:484.

---

the tyrannical yoke of the Stagirite—the famous *Aristoteles dixit*—this did not happen because Aristotle's ideas were known throughout the Middle Ages. On the contrary, in the thirteenth century, Aristotle's theories were a novelty and marked, as it were, the first scientific revolution.

Among the great names of the thirteenth-century revival of Aristotelian science, we should mention, in France, Albertus Magnus (Place Maubert, the square in Paris where he taught, still bears the name of "maître Albert" in shortened form). In England, we should recall Roger Bacon (not to be confused with Chancellor Francis Bacon, politician and philosopher of the seventeenth century). Albertus Magnus was born around 1200; Bacon was some years younger (1214–1294). Albertus Magnus taught Thomas Aquinas (1225–1274), author of *Summa theologiae*. For the fourteenth century we should add Jean Buridan, born ca. 1300, rector of the University of Paris, often remembered for his allegory of the donkey that, placed between water and oats, died of hunger and thirst because it could not choose between the two.

However, Aristotle's ideas penetrated very slowly and counter to religious tradition. In 1215, the Fourth Lateran Council condemned the Stagirite doctrine. Half a century later, in 1277 (as we mentioned earlier), the bishop of Paris attacked such famous followers of Aristotle as Thomas Aquinas and Roger Bacon for their denial that God could have created other worlds. In hindsight, the bishop's argument seems rather bold: in 1600 the unhappy Giordano Bruno was martyred for believing in a plurality of worlds.

## The 36,000-Year Cycle

Now that we have laid, in summary fashion, the historical framework, let us try to visualize the astronomical and geological world of the medieval philosophers. It is often believed today that the six to eight thousand years assigned to the history of the earth by the Bible represented an obstacle to the understanding of geological phenomena—a time that, according to modern science, must be counted in millions of years (the earth, including the solar system, is almost five billion years old). However, this time limit was introduced only in the seventeenth century (see chapter 3).

The real obstacle to the advancement of science was belief in the closed universe: that the world was centered around the earth, and the stars regulated terrestrial phenomena. Today, we know that celestial bodies are in motion. In addition to the earth's daily rotation, which affects the entire sky, there are the movements of the planets, named "wandering stars" because they wander through constellations of stars. The stars themselves, which appear fixed (excluding the earth's daily rotation) when viewed from earth, are actually carried along in a slow circling movement that takes them back to their initial position at the end of approximately twenty-six thousand years. This movement, called precession—explained today by the

---

## Dante's Universe

In the Middle Ages, Dante's (1265–1321) universe in his great epic poem, *The Divine Comedy*, represented the blending of ancient Greek and medieval Christian cosmologies. In *The Divine Comedy* the earth was naturally at the center of the universe, with nine or ten spherical circles of hell inside the earth. At the surface of the earth were two hemispheres: one terrestrial, centered on Jerusalem; the other oceanic, with the mount of purgatory rising out of the vast ocean opposite the holy city. Paradise was reached by passing through the eight celestial spheres of the stars, from the moon (first sphere) to the fixed stars (eighth sphere). These eight celestial spheres were surrounded by the ninth, the *Primum Mobile*, which controlled the whole universe, and the tenth, the Empyrean, God's throne.

---

# The Stars Attract Mountains

At times, astrology took on forms that surprise and amuse the modern reader. For instance, in the thirteenth century, an Italian philosopher, Ristoro d'Arezzo, maintained that the stars attracted the earth "as a magnet has the power to attract iron filings" and produced mountains and valleys. He explained the difference in the size of mountains by the position of the stars. Indeed, stars were known to form the eighth sphere, which had a certain thickness, so that stars (according to their respective positions) were at slightly different distances from the earth. According to Ristoro, stars drew in the sky a landscape "montuoso e valloso" (of mountains and valleys), which was reproduced at the surface of the earth because (contrary to Newtonian gravitational attraction) the highest stars created the highest mountains.*

---

*Frank Dawson Adams, *The Birth and Development of Geological Sciences* (New York: Dover, 1954), 336–338.

---

oscillation of the earth's axis, which can be compared to that of a toy top—had already been observed by the ancients. In the Middle Ages it was thought to last thirty-six thousand years.

Medieval science was permeated with astrology. Sublunar phenomena—i.e., on earth, below the sphere of the moon—could be understood only by the influence of the stars of the superlunar world (situated beyond the sphere of the moon). This is why the ancients' ideas on cyclicity, examined earlier, continued to dominate geological concepts.

The longest cycle then known was the 36,000-year cycle. Every 36,000 years, the stars returned to their original configuration. If the stars regulated terrestrial phenomena, then these phenomena had to abide by the same cycle. This duration, which was long in comparison to the 6,000 years accepted by the people of the seventeenth century (chapter 3), yet ridiculously short with respect to modern knowledge, was the only time constraint on geological phenomena.

## Neptunism

In the tenth century, two theories were pitted against each other in the Arab world. These theories might be named *neptunism* and *plutonism*, adopting the terminology used at the end of the eighteenth century; but since they are applied here to a much earlier period, some reservations must be made.

Neptunism refers to Neptune, Roman god of the sea, and plutonism to Pluto, the god ruling the underworld. According to neptunism, the surface of the earth was formed by sedimentary processes, deposition of sediments in water; whereas plutonism stressed internal processes to explain the earth's topography. These processes, by the way, refer to both Pluto and Vulcan (the Roman god of fire), in particular when the process is volcanic. In that case, one speaks of a vulcanist theory.

Among the neptunists were the Brothers of Purity, authors of the *Encyclopedia*. This collective work was started at the beginning of the tenth century or a little earlier, and finished before A.D. 980. The authors were connected with the Ismailian doctrine, after Ismail, sixth imam (follower of Muhammad). The modern character of this movement lies in the doctrine that proclaimed triumph of the mind over the word and of truth over the law.

The Brothers of Purity taught that erosion destroys mountains continuously. Duhem explained: "Rain waters transport rocks, stones, and sand to the bed of torrents and rivers; those, in turn, carry these materials to swamps, lakes, seas," where they are piled up "in the form of superposed layers . . . ; intermingled, these deposits accumulate one on top the other so that on the seafloor, mounds, hills, and mountains emerge as in the steppes and deserts where the wind blows and sand is piled up in dunes."[3]

The Brothers believed that lands are worn down by erosion and slowly reach the level of the sea, while the hills and mountains deposited on the seafloor cause the level of the water to rise. At a certain moment, the sea overflows its basin and invades the land. This is how periodically—in fact, every thirty-six thousand years—"plains change into seas, whereas seas change into plains and mountains."[4]

With hindsight, something seems quite amiss with this theory. Indeed, sediments deposited on the seafloor would never reach an elevation higher than the surface of the most eroded lands. Hence, even if the lands were completely eroded to peneplains and the ocean completely filled with deposits, the reliefs would never be reversed. The sea would simply cover evenly the smooth surface of the earth.

This was understood by an unknown Islamic author, who wrote in the book *Treatise of Elements* (erroneously attributed to Aristotle by the Christian world) that at this stage all evolution would stop. For the cycle to continue, he said, the ancient sea must rise or the land must subside. The purely neptunistic theory of the Brothers of Purity was thus confronted with one that proposed vertical forces capable of lifting or lowering land.

## Plutonism

Avicenna (ibn Sîna, 980–1037) mentioned such processes in his *De Mineralibus* (On minerals): When "a violent earthquake . . . raises part of the ground . . . a mountain is suddenly formed."[5] Albertus Magnus (ca. 1200–1280) proposed a similar theory two centuries later.[6] According to him the essential cause of mountain building was also seismic. To explain orogenies he suggested that vapors released from the interior of the earth would lift, in their effort to escape, mountainous areas.

These theories show why the term *plutonism* must be used here with great care. As promulgated by James Hutton (1726–1797) in the eighteenth century, plutonism referred to the upwelling of matter in fusion through the earth's crust; but plutonism according to Avicenna and Albertus Magnus did not amount to more than the effects of released vapors. The predecessors of Copernicus and Galileo believed that the earth was formed by the element "earth," the heaviest element in the universe. The four "elements"—earth, water, air, fire—of ancient chemistry were superimposed in that order from the center of the earth toward its surface. Thus, vapors trapped in the earth escaped toward the outer atmosphere. The concept of a central fire, as adopted later by Hutton and before him by Descartes, made no sense in terms of the elements of ancient chemistry. Therefore, we must be cautious with terminology when emphasizing either the evolution of ideas or the discontinuities between concepts in successive periods.

The ideas of Avicenna and Albertus Magnus clearly represent a medieval return to Aristotle's ideas after centuries of neoplatonism. These new ideas were perhaps simply a little more precise, more explicit than those proposed by Aristotle himself, but their meaning was the same. Avicenna and Albertus Magnus were thus an extension of Theophrastus, and their ideas had already been put forth in the writings of Ovid and Strabo.

This is not at all surprising since both ancient and medieval authors viewed the universe in the same way. Furthermore, identical facts served to establish "geological" theories. The remains of living beings had been found in rock layers located far from the sea, so it was concluded that the sea level had once been higher in these areas (or that the ground had been lower). Furthermore, observers had noticed the erosion of mountains and transportation of their debris toward the plains, and from there to the sea; they concluded that, on account of these processes, mountainous lands would finally become level. Since it was believed that the earth was old, if not eternal, it was inferred that mountains had not been formed at the beginning of things—hence the theory of the Brothers of Purity: present lands are ancient seafloors and present seas are future lands.

But even then, there were obstacles to a clear understanding of that theory. Thus, an author as shrewd as Albertus Magnus, who certainly agreed entirely with the concept of an inversion between lands and seas, showed little understanding of the meaning of fossils. After saying that "gluey and viscous mud," which cements the earth into hard rocks, had been transported by water, he added that this is proven by "remains of animals living in water" and ships (still!) found "in the rocks of mountains and in caves."[7]

Albertus Magnus assumed that the waters transporting these remains had come from the sea through subterranean conduits. This theory also dated back to antiquity and had its followers later on. Therefore, it should not bother us. However, it is surprising to hear Albertus asserting that fossils were formed in this cement rather than at the bottom of the ocean.

The main obstacle to an understanding of the transformation of the seafloor into land, however, was not the interpretation of the origin of fossils, but time. Although plenty of time was supposed to be available, it was insufficient to explain these transformations satisfactorily. Time was not scarce in the absolute but it was in a relative sense at the level of the cycle: thirty-six thousand years was too short a time for an inversion between land and sea. The unknown author of the *Treatise of Elements* stated that seashores had changed very little in the previous thousand years and that this fact destroys "the theory according to which the sea was supposed to have changed place at the surface of the Earth." Longer cycles had therefore to be invented. Buridan did just that in the fourteenth century. His view is original and warrants a detailed analysis.

## Billions of Years

In his "Questions about the Treatise on Meteors" (in *Questions about [Aristotle's] Four Books on Heaven and Earth*), Buridan examined the time factor with great care. He first asked: "Was land once in the area where the sea is today and, conversely, was the sea once in the area where land is today and will it return there?"

He then showed that the cycle of thirty-six thousand years was insufficient to answer that question because "if it [the cycle] were the cause of this change, then in eighteen thousand years, the ocean would be in the place where land exists today, and vice versa." However, from observations by Egyptians we know that for the past four thousand years, seashores and mountains have been in the same places where we find them today, whereas according to this cycle "a quarter [of the land] should be covered by the sea."

But, added Buridan, the longest cycle is not the 36,000-year cycle. Indeed, smaller cycles, such as those of the planets, are not mathematically perfect submultiples of the cycle of oscillation (36,000 years) as commonly believed, but are in fact of random duration. Therefore, to find a situation in the sky where all heavenly bodies, including the stars and the planets, would again be found in the same position, it is necessary to use the lowest common multiple of all these cycles. "However, such a situation occurs perhaps only once in hundreds of millions of years." This answers the objection because "if we assumed that during a certain span of time the ocean rotates around the Earth . . . the period of this displacement would be so great—a hundred thousand million years perhaps—that in three or four thousand years, it would be scarcely possible to notice an appreciable part of this displacement."[8]

We should take note briefly, without going into any details, how independent this fourteenth-century scholar at the University of Paris was from the Church.

## The Rotation of the Ocean

Having thus considerably lengthened the duration of the cycle, Buridan argued a series of propositions that explained the slow rotation of lands and seas around the globe. He claimed first that the lands that have emerged are lighter than ocean floors because they have been heated by the sun. Consequently, if one hypothesizes the existence of a continental (thus lighter) hemisphere and of a relatively heavier oceanic hemisphere, the earth's center of gravity would not coincide with its geometrical center. In fact, the center of the world (globe) corresponds to the center of gravity, and the surface of the ocean is a sphere centered upon the center of the world.

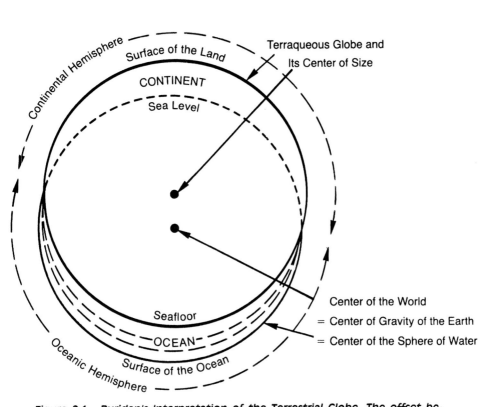

*Figure 2.1. Buridan's Interpretation of the Terrestrial Globe. The offset be-tween the two spheres is highly exaggerated for a clearer picture.*

In other words, Buridan argued that the globe is made up of two slightly offset spherical surfaces. One corresponds to the surface of the land, the other to the surface of the sea. Because of this offset, continents dominate the sea in one hemisphere, while the surface of the ocean is higher than the ground in the oceanic hemisphere. The difference is the distance between the center of gravity (center of the world) and the center of the terraqueous globe (includes both land and water) (fig. 2.1).

Buridan believed that erosion makes continents lighter and bur-dens oceans with the deposit of sediments, and that erosion thus dis-places the center of gravity, which separates itself from the center of the world. Nevertheless, the center of gravity returns to the center of the world in small movements: the eroded lands are being uplifted with respect to sea level, while the seafloor is sinking. Hence, ter-restrial as well as seafloor reliefs are preserved. If changes go on in-definitely, the ancient seafloor should reach the center of the world, while the ancient center should break through the continental hemisphere.

This ingenious hypothesis explains, however, only the preservation of terrestrial reliefs, not the rotation of continents. According to Buridan, erosion does not take place uniformly, and the location of mountains changes with time. The occurrence of springs and rivers depends upon the location of mountains and hence follows them. Therefore, interior seas receive more waters and their level rises. They may thus invade lands and enlarge the ocean at the expense of the continent. If the ocean were thus able to gain ten leagues (about three miles) eastward in ten thousand years, it is possible that in a similar duration of time, it would gain as much toward the east and that the same movement would go on until the ocean has made a complete turn around the earth.

Buridan's concept invites some questions. Indeed, if the ocean gains toward the east (and if symmetrically the continent enlarges toward the west on the other side) at a certain moment, why should this movement not be reversed at another time? If these movements are dependent upon the erosion of land and the filling of marine basins, would not their direction be more or less random? It is possible that Buridan simply wished to get rid of the astrological framework still weighing down his contemporaries. He therefore assumed that, if the stars no longer regulated terrestrial movements, these movements need not be either directional or cyclical. Extremely long cycles therefore became useless. In maintaining them, Buridan was perhaps trying to convince listeners and readers who were still subject to traditional ideas.

## The Asymmetry of the Earth

The relationship between weight reduction (by heat) and the location of land is also unclear in Buridan's theory. If we assume a globe with one continent and one ocean, the two are stable because the lightened continent displaces the center of gravity in relation to the center of the earth. However, if we consider an ocean that surrounds the earth evenly, the two centers would coincide and the globe would also remain in that state. Nevertheless, Buridan maintained that if God "had arranged the ocean as a spherical layer all around the Earth, and if he had allowed the water to flow . . . the water would continuously flow from that part which is now uncovered toward the other part until all of it would be gathered where it is presently."[9]

Nicole Oresme (1325–1382), bishop of Lisieux, Normandy, France, a very learned man and contemporary of Buridan, admitted more logically that the continental part is not "of so much weight" as the other part because it is not formed by "pure earth" but by a mixture of elements.[10] We can understand this asymmetry, he said, if we explain it by the intrinsic density of the globe. The asymmetry is also conceivable if we admit that a hand (of God, naturally, the only being capable of such a gesture) had initially, with a flick of the finger, created a continental hemisphere, because the action of the sun upon this emergent land would have caused precisely such an asymmetry.

Albert of Saxony, another contemporary of Buridan, did not hesitate to propose the following solution: "This asymmetry was preordained by God since eternity for the good of animals and plants."[11] Buridan could not accept this statement and dared to propose that "the world has existed forever as Aristotle seemed to believe even though it is false in the opinion of our faith," a position that took quite some liberties with Scripture for the mid-fourteenth century. Four centuries later, Buffon was to have greater difficulties with the faculty of theology for straying a little from the story of Genesis. He was thought to be extremely bold for maintaining that the "days" of Creation might be simply "long periods" (see chapter 7).[12] Indeed, there was far more freedom in matters of Church doctrine in the fourteenth than in the eighteenth century.

## Leonardo da Vinci on Fossils

There is no doubt that the influence of Buridan and his disciples at the University of Paris extended into the following century. Leonardo da Vinci, born some 150 years later, might have been one of Buridan's followers. A universal inventor and genius as well as a precursor in several sciences, Leonardo is best known as an outstanding artist: the painter of the Mona Lisa and The Last Supper. Born in 1452 in Florence, he died in 1519 in Clos-Lucé, a suburb of Amboise (France) where Francis I had invited him to stay. Historians of science hold Leonardo in respect for his many technical projects and scientific notes of great insight found in his seemingly cryptic manuscripts. These manuscripts, written in mirror script, from right to

# The Origin of Springs

In Renaissance medicine, the human body represented a miniature (microcosm) of the universe (macrocosm). Vice versa, nature was often described through the knowledge of human anatomy. Thus, Leonardo da Vinci wrote about springs:

> It is the property of water that it constitutes the vital human of this arid earth; and the cause which moves it through its ramified veins, against the natural course of heavy matters, is the same property which moves the humours in every species of animal body. But that which crowns our wonder in contemplating it is, that it rises from the utmost depths of the sea to the highest tops of the mountains, and flowing from the opened veins returns to the low seas; then once more, and with extreme swiftness, it mounts again and returns by the same descent. . . . Thus by these two movements combined in a constant circulation, it travels through the veins of the earth.*

He later abandoned the concept of subterranean conduits when he recognized that "waters of rivers come from clouds."

In 1580 the French craftsman and scientist Bernard Palissy in his *Discours admirables,* a dialogue between "Practice" and "Theory," refuted the ancient concept of subterranean conduits:

> PRACTICE. When I had long and closely studied the origin of springs of natural fountains and the place whence they could come, I finally understood that they could not come from or be caused by anything but rains . . .
>
> THEORY. After hearing your statement, I must say that you are a great dolt. Do you think that I am so ignorant as to believe more what you say than in many philosophers who say that all waters come from the sea [implied: by conduits] and return to it?

left, were discovered long after his death and were published only at the end of the eighteenth century, so they could not have influenced his contemporaries. Nevertheless, they bear an interesting testimony to the ideas of a genius who was separated from Copernicus by less than a generation.

Leonardo understood the importance of fossils perhaps better than anybody of his time. Although the presence of fossils on land had been explained since the ancients by the assumption that the seas had dried up, some contemporaries of Leonardo still invoked

(continued)

> p. If I were not so sure of my opinion you would shame me
> greatly: but I am surprised neither by your insults nor your fine
> language: for I am quite certain that I should win against you
> and all those who are of your opinion, even Aristotle and all the
> excellent philosophers of all time . . .
>
> T. Let us come to the proof: give me some reasons which will
> show me that there is some semblance of truth in your opinion.
>
> p. If it were as you say, according to the opinion of the philoso-
> phers, that the springs of fountains should come from the sea,
> it would follow that their waters should be salty like those of
> the sea, and moreover that the sea should be higher than the
> highest mountains . . . and even if the sea were as high as the
> highest mountain, still it would be impossible for mountain
> springs to come from the sea. . . . For the land in many places
> is full of holes, cracks, and abysses, through which the water
> coming from the sea would come out on the plain, through
> the first holes, springs, or abysses it would find, and before it
> should climb to the tops of mountains, all the plains would be
> drowned and covered with water."[†]

---

*The Notebooks of Leonardo da Vinci*, compiled and edited from the original manuscripts by Jean Paul Richter, 2 vols. (New York: Dover Reprint, 1970), 2:197.

[†]Bernard Palissy, *Discours admirables . . .*, 1580, in *Oeuvres complètes* (Paris: Albert Blanchard, 1961). We used here the translation by Aurèle La Rocque, *The Admirable Discourses of Bernard Palissy* (Urbana: University of Illinois Press, 1957), 48–52.

the influence of the stars to argue against the belief that fossils were true animal remains.

Leonardo refuted the stellar theory by showing that the fossil remains were indeed those of animals that had lived and grown because it is possible to count the number of years of growth on their shells. He then declared that these organisms had been deposited at the bottom of the sea, thus opposing the theory that water had gushed out through subterranean conduits.

Leonardo, in fact, was not objecting to the theory of Albertus

Magnus (which he seemed to ignore), but to another very popular idea of the day: the theory that accepted the organic origin of fossils and their formation at the bottom of an ocean but ascribed their unusual presence to the biblical Deluge. Leonardo maintained, "And if the shells had been carried by the muddy deluge they would have been mixed up, and not in regular steps and layers, as we see them now in our time."[13]

Leonardo had reached three correct conclusions: (1) fossils are ancient living organisms and not products of stellar influence; (2) they have been deposited on the seafloor; and (3) the sea has been there a long time, certainly more than forty days. Albertus Magnus had not gone beyond the first conclusion. The same is true for Bernard Palissy (ca. 1510–1589), the potter from Charente often regarded as Leonardo's successor. Palissy at first assumed, rather naively, that fossil shells were leftovers from some traveler's meal, later petrified in the earth by means of "some salsitive seeds." Indeed, almost two centuries later, Voltaire (1694–1778) was harshly criticized by Buffon for a similar proposal, namely, that "rotten fish were thrown away by a traveler and were petrified thereafter."[14] In an attempt to go beyond this modest hypothesis, Palissy attributed the accumulation of shells in the coarse limestones of Paris to "great lakes" or "trenches, cavities, and receptacles of water which entered into the cracks of these rocks."[15] This is certainly very inferior to Leonardo's intuitions.

## Collapse of a Cavern beneath the Seafloor

It remained to be explained how the seafloor emerged. It is possible that Leonardo knew about the ideas of Buridan and the school of Paris because he said, "The softened side of the earth continuously rises and the antipodes sink closer to the center of the earth."[16] Pierre Duhem, the early twentieth-century French physicist and historian of medieval science, argued that Leonardo had borrowed a great deal from the school of Paris of the fourteenth century. Indeed, Buridan's ideas were certainly still discussed at the end of the fifteenth century. Hence, until Duhem exhumed the manuscripts of Buridan, Leonardo's ideas in his manuscripts looked even more original than they really were.

However, Leonardo explained uplift of the ancient seafloor quite differently from Buridan, a fact that historians of science seem to have overlooked although it is perhaps Leonardo's most original contribution. I mentioned earlier that Buridan's theory about the rotation of the ocean and the land has its shortcomings. Leonardo did not mention this cycle at all. To explain that "the summits of mountains tower so high above the sphere of the water," he imagined rather that a huge cavern beneath the seafloor, filled with water, must have collapsed, drained by the flow of subterranean conduits toward the center of the earth. The crust above the collapsed cavern was "uplifted" because it "was lightened by this [collapsed] part and even further by the weight [loss] of the waters above." This is how "marine shells and oysters found on high mountains" can be explained.[17]

This idea is ingenious. Until then, naturalists had still tried to understand the asymmetry of the globe, which supposedly caused the waters to be higher above the land in one hemisphere and the reverse in the other hemisphere. Contrary to Albert of Saxony's dictum that the earth was regulated by God, Leonardo preferred to find a natural cause that explained in a very simple fashion the presence of fossils on land. In fact, he believed that the sea covered the entire globe before this catastrophe. Hence, it was no longer necessary to exchange lands and seas to understand the presence of fossils in mountains, as all theories of cyclicity proposed. Leonardo replaced cyclical reversals by a unique collapse, some sort of historical event—an idea that anticipated Descartes' concept, to be examined in the next chapter. It is this idea that makes Leonardo seem so modern. Indeed, although Buridan's ingenious theory that the continents are uplifted thanks to erosion while the seas are sinking under the weight of sediments brings to the minds of modern geologists the idea later named **isostasy,** Buridan still invoked the processes of a pre-Copernican cosmology in which the earth was still the center of the universe. Leonardo—who wrote in one of his notebooks that "the sun does not move"—proposed a theory that maintained its power of interpretation long after Copernicus: in the seventeenth and eighteenth century, subterranean collapse became the general explanation for mountain building.

# How the Earth Was Formed

## Nicholas Copernicus

Nicholas Copernicus, a Polish subject, died in 1543. Twelve years earlier, he had finished his manuscript *On the Revolutions of the Heavenly Spheres* (in Latin, *De Revolutionibus* . . .), a truly revolutionary work in the modern sense of the word. The book was published shortly before he passed away, and it is said that he received the first copy on his deathbed.

A. Osiander, the somewhat heterodox Lutheran disciple who saw the work through the press, must have immediately understood that the Copernican theses were of an explosive nature because he felt the urge to add an anonymous preface, long attributed to Copernicus himself. He explained that astronomers must not search for hidden causes nor for the real movements of heavenly bodies. They should simply "save the appearances" (in the literal sense) and present hypotheses that describe these movements as precisely as possible. Osiander added, "Hypotheses are not articles of faith but merely bases for calculations."[1]

With these premises, the new astronomy could easily show its superiority over the Ptolemaic system. The Ptolemaic system—now more than a thousand years old—had become increasingly complicated; new observations forced astronomers to revise and adapt to the system, so that its use became more difficult day after day. Legend tells us that a king of Castile in the thirteenth century exclaimed that the Almighty should have consulted him before Creation because he would have proposed something simpler. Copernicus' bases for mathematical calculations were thus accepted without great hesitation. In 1582 the Catholic church decided to reform the calendar because the length of a year calculated under Julius Caesar was slightly wrong; after sixteen centuries, the consequences had become important. Copernicus' system was used for this correction. Pope Gregory XIII did not seem to mind that he owed his calendar to Copernicus, any more than his predecessor, Pope Paul III, had objected to having De Revolutionibus dedicated to him.

However, Protestants, who recognized the Bible as the only authority and who believed in a literal interpretation of the Scriptures of Christianity, did not appreciate the ideas of "this fool who wishes to reverse the entire science of astronomy because," said Luther, "Sacred Scripture tells us that Joshua commanded the sun to stand still, and not the Earth." Poking fun at Protestant dogma was risky, as the unhappy fate of Michael Servetus proved: while taking refuge in Calvin's city, he spoke ill of the Trinity and was burned at the stake for it. The Catholic church also was forced to counter the Copernican system when it met at the Council of Trent in 1545 and decided upon a repressive program.

## Galileo

Problems of astronomy were in fact not at the heart of the ensuing controversy between Catholics and Protestants. The controversy began with Galileo (1564–1642). When he pointed his telescope (his own invention) at the sky, the Italian physicist discovered multiple proofs that the Copernican system not only made it easier to calculate the movements of the stars, but also corresponded to reality. Johannes Kepler (1571–1630) presented similar arguments independently. This time, the Catholic church felt the threat and reacted. In 1616 the Holy Office condemned Galileo, and the De Revolutionibus was put on the Index, the list of books that the Roman Catholic church condemned as dangerous to faith. However, Galileo neither gave in nor lost his confidence. Once his friend Barberini became pope (1623), under the name of Urban VIII, Galileo felt comfortable enough to publish his Dialogue Concerning the Two Chief World Systems in 1632. However, because the dialogue caricatured a traditional thesis, defended by the character of the naive Simplicio (whom some wicked minds suspected to represent the pope himself), the *Dialogue* led in 1634 to Galileo's trial in which he was condemned and forced to recant his errors.

We may be surprised at the intransigeance of the Church in its defense of the Ptolemaic system if we recall how cautiously the first Christians had treated it. Although the Ptolemaic universe of concentric spheres stood in contradiction to the naive model of Genesis, it was by this time part of Church teaching: former disdain for the pagan science had all but vanished. So, critiques of the system that had lasted for so many centuries felt like direct attacks upon the Church itself.

More important, the Church was perhaps aware that the emerging science was contesting its authority more than its knowledge. In the thirteenth century, when the bishop of Paris defended the principle of the plurality of worlds against Thomas Aquinas, arguments on both sides were purely theological: Is God's power restricted if we assume that the world is unique? Is it not a blasphemy instead to maintain that God created several worlds? In fact, God must have started several times if some of the worlds are not perfect, or must have stuttered if they are of equal perfection. The dilemma belonged thus to religion alone, and the Church could rest assured that it would have the last word.

However, with the new science, arguments were based on reason and observation, which claimed authority by their objectivity. The Church was in collision with a powerful rival and felt the pressure to find in scientific facts "proofs" of the truth of the biblical narrative on the creation of the world. From then on, the Church was going to need the flexibility of Christian scientists who would have to put their observations at the service of their faith. Most certainly, this problem prompted the Church's efforts to condemn Galileo. But the condemnation was in vain; the new science continued to advance.

## Descartes

When Rome condemned Galileo, René Descartes, a French philosopher living in Holland, was ready to publish *The World, or a Treatise on Light* (in French, *Traité de la lumière*), in which the final word on light, stars, the earth, and humans was to be presented. Descartes was born in 1596, during the reign of Henry IV, in a village in Touraine later named after him (La Haye-Descartes). He was a sickly child, and later on a student at the Jesuit college of La Flèche before starting to travel, at first with the army and then on his own. He settled in Holland because he was afraid of losing his freedom. He wrote to a friend: "What other country is there where one can enjoy such complete freedom"?[2] Descartes had heard that the poet Théophile de Viau, whom he respected, had been exiled, and he knew that Lucilio Vanini, Italian philosopher, had been burned at the stake in Toulouse for being an atheist. Descartes was not ready for such a sacrifice.

When he learned about Galileo's misfortune in 1634, he halted publication of his treatise. However, he did not remain completely silent and published in 1637 the famous *Discours de la méthode* (Discourse on method), followed by scientific essays on *Dioptrique* (Optics), *Météores* (Atmospheric phenomena), and *Géométrie* (Geometry). After his 1641 *Méditations philosophiques* (Meditations on first philosophy), he published *Principes de la philosophie* (Principles of philosophy), first in Latin, then in French (1644, 1647), where he gave his opinion on the formation of the world, which we shall analyze in this chapter.

After writing *Traité des passions* (The passions of the soul) and *Traité de l'homme* (Treatise on the human body), published after his death, Descartes left Holland, where he had begun to feel persecuted, and went to Sweden. Unfortunately, its climate was not good for his health, in particular when Queen Christina, who was very demanding of her philosopher, summoned him to her palace at dawn. He died shortly after his arrival in 1650.

In his *Principles of Philosophy*, Descartes cleverly explained that he "denied the movement of the Earth with greater care than Copernicus."[3] Indeed, he said that the earth does not move in the sky, but "it is nevertheless being carried by it." He compared the earth to a ship on the sea that is propelled neither by the wind nor by oars, but moves imperceptibly back and forth with the movements of tides. He thus rejected the commonsense arguments of medieval authors who had said: If the earth really moved, birds would not move at the same speed when flying in the direction of the rotation or in the opposite one.

The Cartesian thesis is bolder than that of Copernicus. Indeed, the Polish astronomer had unbolted the earth from the center of the universe only to replace it by the sun, so that his system still remained hierarchical and centered. Earth simply took a modest rank among planets; the stellar sphere, however, remained centered upon the same spot as the lower spheres, although Copernicus admitted that the stellar sphere had an infinitely larger radius in comparison with that of the earth.

Descartes further noted, "It is necessary to assume that fixed stars are extremely far away from Saturn." He added, however, "that the stars are not all located on a spherical surface but are far away one from the other."[4] With these words, the world was opened and the closed universe of the ancients collapsed. All hierarchy among parts thus vanished. To this new astronomy corresponded a new physics where "upward" and "downward" were no longer absolute directions.

## The Earth Is a Sun

I have strayed far from geology and must reassure the reader that this is the last time that astronomy shall be referred to. Once removed from its central position, the earth became autonomous and could be studied for itself. Its position in the center of the world had placed it in an embarrassing and ambiguous situation. Of course, it had been, so to speak, in the limelight. (By the way, one of Copernicus' arguments for replacing the earth with the sun was that the sun is more beautiful.) But the earth had also included hell. Indeed, according to the hierarchical world of Aristotle's physics, the most noble elements occurred at the periphery of the cosmos. The stars above were perfect, unchangeable, and incorruptible, but the sublunar world was one of corruption. Finally, the element earth was the most inert of the four elements. The observation of a nova (stellar explosion) in 1604 was instrumental in the acceptance of the new astronomy because it implied that the stars were not perfect.

In short, when earth lost its privileged place, it also relinquished its status as a vile body. Descartes changed the earth into some sort of abortive sun. It came closer to the status of stars and acquired a central fire.

Furthermore, a hierarchical, finite universe had required maintenance of its structure and repair in case of degradation. Indeed, the Aristotelian earth maintained a dynamic stability through the action of partially interacting cycles. However, if the earth was a mere speck of dust in an infinite universe, its birth and origin became of interest, regardless of the fact that it may disappear or may not have existed forever: the earth had the right to a personal history.

It is this history that Descartes started to narrate in the fourth part of his *Principles of Philosophy*, after having described the structure of the world and the formation of the sun. In order to explain the composition of our planet according to Descartes, I shall briefly mention some essential points of his work.

The Cartesian universe consisted of three kinds of substances. Initially, it was formed of irregularly adjoining particles (because the Cartesian world had no vacuum). However, rubbing against each other, these particles became rounded and lost scrapings of smaller size. These scrapings had a tendency to aggregate and to form in part a *third* type of substance, whereas the nonaggregated "scrapings" formed the *first* substance and the rounded particles the *second* one. In essence, the first substance (of nonaggregated scrapings), which was transmitted easily through the second one, was the matter of light; the second (rounded particles), which was transparent, was the matter of the sky; and the third (aggregated scrapings), which blocked light, formed terrestrial bodies.[5]

Matter organized itself into vortexes because in Descartes' concept of absolute fullness, unlike the vacuum of atomists, movement is not possible in a straight line.[6] Thereafter, vortexes organized themselves internally. Matter of the first type concentrated in the center, forming a star. It became covered with spots, which consisted of substances of the third type: these were sunspots, which, when thickened, formed a crust and thus changed the star into a planet.

Planets, born in smaller vortexes than the sun, moved downward toward the sun after having captured their satellites, and the fourteen vortexes of the "first sky" formed our present solar system.[7]

## Formation of the Earth

According to Descartes (see fig. 3.1, a and b), a layer (*M*) formed around the central fire (*I*), a layer similar to sunspots "filled with a very opaque, or *dark*, and very *solid* and dense body." The peripheral zone *B*, consisting of loose matter of the third type, surrounded the layer *M*. Zone *B* differentiated progressively while the earth moved downward toward the sun because it encountered particles of the second element, which were larger than its own and prompted its matter to condense. Thus was formed zone *C* around *M*. The area *D*

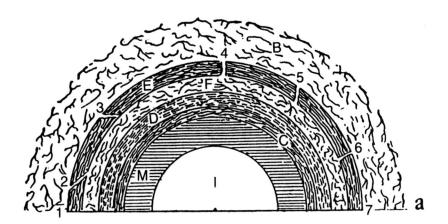

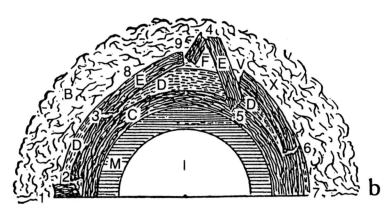

LEGEND: I = central fire; M = layer of the material of sunspots; B = atmosphere; C = lower earth's crust (metallic); D = ocean; E = upper earth's crust (stony); F = subterraneous cavity

*Figure 3.1. The Formation of the Earth (Modified after Descartes,* Principles of Philosophy, 1644*). The two diagrams (a and b) show the last two successive stages of the formation of the earth.*

was separated from C in the same fashion as water is squeezed out from a swamp when one walks on it. A new zone of solid particles was deposited around body D, called area E. Finally, the area E separated from D by interposition of body F (fig. 3.1a).[8]

Descartes explained that "B and F are nothing else but air, that D is water, and that C is an internal, very solid and *very heavy* earth crust from where all the metals are extracted, and finally, that E is *another earth crust, less massive,* which consists of stones, clay, sand and mud."

---

# Vegetation of Stones

Descartes took the opposite view of the analogy between macrocosm and microcosm; he wanted to reduce the animal to a machine (theory of the animal machine). However, this "machine" did not convince all his contemporaries; many of them continued to adhere to anthropomorphic or animistic ideas. The analogy between the universe and the human body was only one of these ideas. Other varieties existed, in particular, the belief in "vegetation of stones."

Miners believed for a long time that stones "grew" and slowly filled in the effects of excavation. Medieval lapidaries talked about the most extravagant ideas on this subject, for instance, of the existence in gold mines of males and females who "nourished themselves from the dew of the sky, who conceived, sired, and gave birth to little ones who multiplied and grew every year."*

In the beginning of the seventeenth century lived Campanella from Calabria (1568–1639), a bold mind whose political ideas were far ahead of his time and who thus spent more than one third of his life in prison. In science he tried in vain to reconcile the new sciences with medieval superstitions such as the analogy of microcosm-macrocosm as well as the vegetation of stones. He saw a proof of this vegetation "in the fact that crystals develop according to specific shapes as animals do according to their species."†

The vegetation of stones fable was repeated even in the eighteenth century. We find it in *Histoire naturelle de l'univers* (Natural history of the universe) by F.-M. Pompeo Colonna, an Italian physician who died in 1726. He wrote:

It seems to me that mountains vegetate and that some of those that we see today have vegetated on Earth like trees. . . . What

---

This globe (fig. 3.1a) had a perfectly smooth surface, an interior ocean ($D$), and an interior atmosphere ($F$). This very unstable situation could not last because crust $E$ was cracked at various places (fig. 3.1a at points 2, 3, 4, 5, 6) "in the same manner and for the same reasons as it commonly happens in the soil of *marshland* when summer heat dries it out." Thus undermined, body $E$ collapsed through less dense zones. "But since this surface was not large enough to receive all the parts of this body in the same situation as they were before, it was necessary that some of them fell sidewise, leaning one against the other." In other words, the collapsing body $E$ had a larger circumference than body $C$, on which it fell, so it could not cover it evenly but had in part to take an oblique position.

*(continued)*

may perhaps surprise the observer is the fact that it is not easy to understand how a rock may grow and vegetate in this manner because we do not see it in the process of being formed but only when it is already formed. Similarly, a person who sees a fragment of coral which is so hard and shiny might be astonished . . . if told that this is a plant which vegetates in the sea. The same would happen to observers of a completely formed mountain if they were told that the mountain became this high merely by vegetating as a tree. We must therefore understand that vegetation of a stone or of a mountain occurs by means of moisture mixed with terrestrial parts which are pushed by heat, that heat moves both matters, and that when moisture evaporates, a hard body remains which does not vegetate anymore.[‡]

The idea of vegetation of stones is the animistic form of the thesis of equilibrium in nature—previously encountered in Aristotle and found again in Lyell—a thesis that was given a more rational form by these authors. Its antithesis stated that the world decays slowly (Stoic and Platonic thesis) and can only be repaired by a revolution.

---

[*]Frank Dawson Adams, *The Birth and Development of Geological Sciences* (New York: Dover, 1954), 96–97.

[†]P. Lenoble, *La géologie au milieu du XVIIᵉ siècle*, Conférence, Palais de la Découverte, Paris (Alençon: Imprimerie alençonnaise, 1954), 10.

[‡]F.M.P. Colonna, *Histoire naturelle de l'univers*, 4 vols. (Paris: A. Cailleau, 1734), 1:204–205.

This is how Descartes explained the formation of mountains, continents, and oceans in his famous figure (our fig. 3.1b). From left to right, we see on the half sphere, an ocean (body $D$, or 2–3); then a continent in an oblique position ($E$, 8–9) with a central mountain (4); then, to the right of the mountain, another oblique continent ($E$, or 4–5); and then again an ocean ($D$). The Cartesian model of the formation of the earth warrants several comments.

First, the Cartesian globe had a central fire of the same nature as that in stars. It had acquired an elevated rank, which the geocentric earth had not. Furthermore, although the central fire was tightly enclosed and showed, according to Descartes, no activity, it harbored a potential energy ready to be awakened.

Second, the morphology of the earth's surface resulted from the collapse of ocean bottoms and plains and not from the uplifting of mountains (the modern interpretation). This is an important point because most authors in the following two centuries adopted Descartes' view.

Third, in his description of the changes of the earth, Descartes put himself in direct opposition to religious cosmogonies, which until then had enjoyed the privilege of saying how the world had been formed. Only Lucretius, seventeen centuries earlier, had dared to propose a rational theory of the formation of the universe. Although the models of the two authors differed because the atomist believed in vacuum where Descartes denied that idea vigorously, both philosophers shared the same ambition.

Descartes' ideas were certainly considered heretical in France in the 1640s. This is why he was careful to emphasize—without much conviction—that he did "not doubt that the world has been created at the beginning with as much perfection as it shows today." But he made no effort to reconcile his system with Genesis. Nevertheless, Descartes' idea of a crust broken by collapse was accepted eagerly by diluvialists (authors who accepted the biblical Deluge in their theories of the earth) because it implied an internal ocean and internal air within mountains. This concept provided explanations for aspects of the Flood, namely, water surging out from the abysses. Thirty years later, the English diluvialist Thomas Burnet followed Descartes and attempted to explain the formation of mountains and plains by collapse, which he then related to the surging out of waters from the abysses mentioned in Noah's Flood. His work was published in 1681 under the title *Telluris Theoria Sacra* (Sacred theory of the earth). Burnet introduced a term, "theory of the earth," which was to be in use for more than a century.

# The Work of God

## The World Is 6,000 Years Old

Thomas Burnet (1635–1715), chaplain and secretary of King William III of England, was not lacking in boldness when he said that he was comparing similarities in the Bible and the Cartesian system on the formation of the world. Descartes was, in fact, saving face when he stated that his story was purely hypothetical and not intended as a substitute for the biblical account. To identify an act of God with a purely natural process subject to natural laws was certainly much more daring. Jacques Roger has pointed out that "eighteenth-century England offered freedom, almost unknown on the continent, for those who took part in the interpretation of the Bible in a too personal fashion."[1]

Burnet simplified the Cartesian system and modified it noticeably. Elements were arranged in order of decreasing density: earth (not fire) was at the center, surrounded by water, an oily liquid, and air. However, dust from the cleansing of the primeval atmosphere settled gradually into the oily liquid and formed a crust which, when dried out by the sun, collapsed into the water below.

What truly separates Burnet from Descartes is his statement, "The Explication we have given of the Universal Deluge is not an Idea only, but an account of what really came to pass in this Earth, and the true Explication of Noah's Flood."[2] He believed that, if God was responsible for anything in this matter, it was by his laws imposed upon the world, but not by his personal intervention; that is, God did not take an action to change the course of nature in order to punish human disobedience.

However, Burnet remained close to the Bible in regard to duration. He limited to sixteen centuries the time separating the first induration from the breaking up of the silty crust. Here we can call this a short period of time, particularly compared to Buridan's millions of centuries. It is a paradox that medieval authors assumed immense cycles, only to have oceans and continents turn with an imperceptible movement around the earth; whereas Descartes and Burnet, who allowed only very short time spans, had the ambition to create the world. Perhaps we must go to the bottom of the paradox to find the answer and to ask whether it was necessary to return to short time scales to sketch a history of the earth. Indeed, this history was first of all the history of humanity. If anyone dared attempt writing this particular history, its entire scope had to be encompassed by a look backwards. In other words, people turned toward their past only when they were able to embrace it at a glance.

## Return to the Origin

It was precisely at the beginning of the sixteenth century that people began trying to trace their origins. The Protestants first read the Bible closely in order to find reasons to criticize the Catholic church. Once the Catholics were accused of having broken with tradition, they took their turn to join in the battle. Initiated by religious rivalry, interest in history began to spread. In France, manuals of history became numerous in the sixteenth century. The *Chronique* (1532) by Jean Carion, for instance, was adopted later on by Philip Melanchthon (1497–1560), a follower of Martin Luther. Melanchthon began to teach history in 1518 and founded a chair of history at the German university of Heidelberg in 1558.

The history of humanity allowed not only a return to its origin but also the dating of the Creation of the world. Because Creation was believed to have happened rather recently, historians took pride in calculating it as precisely as possible. Today, we may find somewhat ridiculous the illusory precision with which pious exegetes tried to determine the age of the universe. In the middle of the seventeenth century, Archbishop Ussher (1581–1656) proclaimed that heaven and earth were created the night before October 23 in the year 4004 B.C.[3] Although not everybody was so precise, the proposed dates varied little: generally, 4,000 to 5,000 years were counted from the dawn of the universe to the birth of Christ. Christopher Columbus was rather generous in reckoning 5,343 years between them. But he was less so about the date of the Last Judgment: he estimated that the millenium (the time separating the Creation from the Last Judgment) covered seven thousand years, so the world had only 155 years left to exist.

It seems that most exegetes believed in an impending end of the world and therefore imagined its creation of little antiquity. Indeed, every philosopher said that the world declined at a breathtaking speed and was therefore not going to last very long. According to the historian Jean Delumeau, Martin Luther would have been less intolerant had he not believed in an impending end of the world.[4] But since decay was so rapid, the world could not be very old, and hence both its youth and imminent death were deduced from the speed of its (assumed) decline. Signs of decay were in fact everywhere obvious to the people of the beginning of the classical age as well as to their predecessors. Tragic events had troubled the Western world since the middle of the fourteenth century.

## The Fear of Mountains

Although fear of the end of the world supposedly lessened toward 1660, short estimates of the age of the world persisted for a long time. Ussher's calculation, the most famous, was included in official versions of the Bible until the beginning of the nineteenth century, when it became an obstacle to the progress of science. However, toward 1680, such short chronologies were actually substitutes for the endless cycles of the ancients and represented something of a prerequisite for attempts at any history of the earth.

Theories of the earth began to appear at the end of the seventeenth century. Nevertheless, there were various ways to construct a theory of the earth, and Burnet's version was perhaps typical of his time. It was believed that, if mountains dated from the Deluge, they must be an obstacle placed there by God so that people could not communicate with one another. Mountains were also regarded as a disfigurement of the surface of the earth; many contemporaries compared them to boils, pimples, or blisters. In 1646 an English traveler wrote

# The Age of the World

Between 1350 and 1660, it was believed that the world was very very old because it had aged very fast. Its creation therefore went back only a few thousand years. The proof of this accelerated and premature aging lay in the many signs of decrepitude witnessed by history.

The first sign was a plague that swept down savagely upon Europe in the middle of the fourteenth century. Coming from Asia in 1346, the plague progressively invaded the Western world. It lasted until 1453 and killed twenty-five million Europeans. After the epidemic subsided, the disease remained in an endemic state with sporadic outbreaks. The plague broke out in Venice in 1478, spared the city the following century because of a sanitary cordon, but ravaged the rest of Italy. In 1655 London suffered from a severe epidemic, and in 1720 Marseille experienced the same unfortunate fate. The plague disappeared from Europe only in the middle of the nineteenth century.

The second tragic event was the great schism in the Western church. Although it destroyed fewer lives, it seemed a striking portent of the dissolution of religious institutions and the end of the world. In 1378 the conclave of cardinals elected a pope and then proclaimed its decision null in order to elect a second one. Two men thus occupied the pontifical throne, one at Rome and the other at Avignon (where the papacy had moved in 1309). An attempt at solving the problem resulted, in 1409, in the proclamation of a third pope. The conflict ended only in 1415 at the Council of Constance where Martin V was elected. Nevertheless, the papacy was weakened, and the council took advantage of this weakness to assure for itself superiority over papal authority.

that "nature has swept all the rubbish on Earth into the Alps," the summits of which look "strange, horrible, and frightful." And, like Descartes at that same time, he thought that mountains "had shattered and collapsed one on top of the other which was rather terrifying."[5]

Added to political and religious signs of human decadence came the testimony of travelers brave enough to overcome the fear of mountains. Indeed, God's wrath was shown not only by the frightful epidemics he had inflicted upon civilization, but also in the way he had disfigured and endangered parts of the earth's surface. The Bible said that the wrath of God originated with the Fall of Adam and Eve and that the Deluge was the most spectacular example of that wrath after their expulsion from paradise. It seemed obvious that mountains were formed during that event.

*(continued)*

The Hundred Years War (1337–1453), the folly of King Charles VI, and the battles between the Armagnac and the Burgundy armies accentuated in France the feeling of an impending end of the world.

Furthermore, at the beginning of the sixteenth century, the Church went through a great period of protest. In 1517, Martin Luther decried the practice of buying indulgences as a lie that provided illusory security because human acts are powerless in the face of God's will. The resulting movement (the Protestant Reformation) became explosive, and religious wars caused bloodshed throughout the century in most European countries, ending in France with the Saint Bartholomew's Day Massacre of 1572 (Paris) and even later in other European countries.

According to Jean Delumeau, publications of works on "the age of the world" and on the "Last Judgment" multiplied during these troubled years. The *Apocalypse* of Albrecht Dürer dates from 1498; and the *Last Judgment* in the Sistine Chapel, the famous painting of Michelangelo, was undertaken forty years later (1536). Even more strange, the first known description of the terrors of the year 1000 was made public at the end of the fifteenth century. One might therefore wonder if these terrors were not perhaps at least partly a creation of that time because eschatological fears could now be circulated with the help of the printing press, which did not exist before the fifteenth century.*

---

* J. Delumeau, *La Peur en Occident (XIVᵉ–XVIIIᵉ siècles): Une cité assiégée* (Paris: Fayard, 1978), 189.

Purely theological objections to Burnet's theory by his contemporaries may be summed up as follows. According to Moses, the Deluge was an exceptional rainfall. He did not mention any gigantic collapses, only the rupture of reservoirs of the "great abyss." Moreover, careful reading of the Bible shows that mountains existed before the Deluge, as witnessed by the olive tree on Mount Ararat; whereas Burnet maintained that the earth's crust before the Deluge was smooth and without any mountains.

## The Use of Mountains

In opposition to Burnet's view of the earth in ruins, some naturalists presented arguments in favor of the physical necessity of mountains. Mountains were considered necessary for the natural water cycle, and the springs of rivers that carry the waters toward the ocean are found mainly in mountainous areas. Whether these springs are fed by subterranean conduits, as believed by the ancients, or whether the importance of evaporation (evaporation of the ocean, which returns the water to the mountains in a cycle) was recognized, as maintained by Palissy and as demonstrated definitely in 1674 by Pierre Perrault,[6] it seemed clear that this natural water cycle would not exist if the earth did not have any mountains.

The first author to emphasize the use of mountains was the English naturalist John Ray (1627–1705). Known by historians for his works on taxonomy, which place him among the immediate predecessors of Carl Linnaeus (1707–1778) for the classification of living beings, Ray was also the author of *Three Physico-Theological Discourses*, a work dealing, among other subjects, with the formation and use of mountains.[7] His reasons for the use of mountains were abundant: their slopes are good construction sites for houses; they offer a variety of climates suitable for various animals and numerous plants; they are favorable for the exploitation of metal and mineral deposits; they grow plants that feed animals; they produce springs of rivers and fountains; and they act as natural boundaries.

In the course of the next century, this kind of argument became a cliché. Noël-Antoine Pluche, author of *Le Spectacle de la nature*, a bestseller on natural history published in the middle of the eighteenth century and reprinted for many decades, showed that the universe was a "wonderful machine."[8]

The author who pointed out the beauty and many uses of mountains perhaps more eloquently than any other was Élie Bertrand, a pastor from Bern, Switzerland. In his 1754 "Essai sur les usages des montagnes" (Essay on the use of mountains), he gave even more details than John Ray. He stated at first that mountains are beautiful and that only superficial minds could have seen in them "debris without proportion, ruins without order, or products of randomness." Furthermore, mountains serve as foundations for the earth and help to preserve it. They enlarge the surface of the earth and act as boundaries between nations. They produce winds and springs and distribute waters. Finally, they contain caves, which provide refuge to persecuted Christians (a seasonable thought during religious wars).[9]

The argument about foundations is partly reminiscent of the analogy between microcosm and macrocosm: mountains form the bones of the globe. However, Bertrand's argument includes also a more modern observation on the disposition of "reentrant and salient angles" of mountains, a concept introduced in Louis Bourguet's theory of the earth, given at the end of *Lettres philosophiques.*[10]

Bourguet said that "mountains are shaped more or less like fortifications." If we observe two parallel mountains, we see that each one juts out in a "salient" toward the other and that the "salient angles" of the first "correspond respectively to the reentrant angles" of the second.[11] Translated into modern terms, this "discovery" in which Bourguet saw nothing less than the "principal key to the theory of the Earth," loses some luster because it simply means that the two flanks of a valley are approximately parallel. If one forms a spur, the other forms a cirque that is molded around the spur. But as long as the origin of valleys was not attributed to erosion by water, the parallelism of their flanks could not be understood. Today, we would not talk about *two* opposite mountains and their angles but about *a* mountainous area cut by an entrenched valley with a meandering river. At any rate, according to Bourguet, the observation of "corresponding angles" attested to the idea of *order* in nature and implied that morphology in mountains responded to a design.

## The Role of Providence

The interest shown in the eighteenth century in the beauty and the use of mountains may be related to their gradual discovery as the fear of mountains slowly disappeared. In the seventeenth century, they were feared (as much as the ocean or the devil and those then regarded as his agents: worshipers of idols, Moslems, Jews, and women), and nobody would normally venture there. High mountains inspired great fear, and the names of some summits—such as Mont Maudit (cursed mountain)—recall this aversion. In the eighteenth century, some fearless travelers entered these mountains and were struck with admiration by the spectacle of primitive nature. In 1739, an English traveler wrote from Turin: "There is no abyss, torrent, or cliff which is not imbued with religion and poetry. Some gorges exist whose terrible majesty would turn an atheist into a believer without the help of any other arguments."[12] In one century the general view had reversed: mountains were no longer heaps of rubbish but admirable proofs of the existence of God.

In his youth, Albrecht von Haller (1708–1777) from Bern, Switzerland, undertook an Alpine journey and published a poem on the

Alps in 1732. He was certainly not the first to see the Alps, but he was the first to describe them. This poem became famous and was translated into French in 1750 and into English in 1794.[13] In 1754 the brothers Jean-André and Guillaume-Antoine Deluc (whose important role in regard to the emergence of modern geology shall be mentioned in chapter 10) went all the way to Chamonix, at the foot of Mont Blanc. In 1760 Horace-Bénédict de Saussure posted a handsome reward at Chamonix for the first person to reach the top of that mountain.

The taste for mountains was perhaps awakened by the many discourses on the general importance of landscapes written by authors influenced by religious considerations. Gordon L. Davies attributed this change in perspective about mountains to a change of religious sentiment:

> The Calvinists had seen God as an awesome, wrathful Being, inflicting terrible punishments upon a sinful mankind, but the faithful of the new era saw Him as the benign and merciful Being who is reflected in the enlightened teaching of the Cambridge Platonists and the Latitudinarians. Thus by 1700, God the angry judge of mankind had given way to God the gracious architect of a magnificent creation.[14]

The Protestants and the Anglicans believed more readily in divine providence, whereas the Catholics were less outspoken in that respect. Of course, Pluche was struck with admiration by the spectacle of nature, as were other Catholic naturalists. However, he differed from them by the role he attributed to the Deluge in the formation of the earth's topography.

## Deluge or Creation?

Believers in divine providence maintained that if mountains were so important for humans to live on earth, they could not have been formed by the Deluge. They must therefore date back to Creation.

John Ray believed that both direct action of God's will and "secondary causes" were responsible for the formation of mountains. Although his argument might have been purely opportunistic (Ray showed that subterranean fires could lift mountains all by themselves), the proposed mechanism was less elaborate than in Descartes or in Burnet. Ray believed that, if mountains dated back to Creation, their origin would be lost amongst that of the rest of the world.

Appealing to arguments about harmony and the use of mountains, Élie Bertrand rejected all theories proposing that mountains were formed gradually or by chance. According to him, the world is far from consisting of ruins, as maintained by Burnet; it can be compared to a well-constructed and beautifully decorated house.[15]

Bourguet, who had emphasized order in the shape of mountains, could also not consider that their formation was due to chance. Although he did not totally exclude the effects of the Deluge, his theory opposed Burnet's theory of "a collapse of the former world"; he nevertheless accepted "the dissolution of the first world." This "second formation" was justified by his observation—which he believed to be as fundamental as that of the correspondence of angles—that fossils are filled with "the same material as that found in the beds and layers where they are buried."[16] According to Bourguet, this situation could be explained only by a dissolution of the world, which allowed fossils to penetrate these rock formations. This dissolution, which Bourguet identified with the Deluge, had occurred sixteen centuries after Creation. The earth had lost its shape at the time of the spring equinox and regained it at the time of the fall equinox. (This statement certainly does not lack precision.)

Bourguet was, however, more vague than Burnet about the processes leading to dissolution and a new formation of the world. He had to admit that the first globe must have been less solid than the present one because of the continuous effect of gravity. Buffon, after Bourguet, was to accept the same theory. Fontenelle, the perpetual secretary of the Academy of Sciences, who summed up regularly in *Histoire de l'Académie Royale des sciences, Paris* all the memoirs received by the academy, wrote as early as 1718 that the globe had suffered "during the first periods of time after its formation, extraordinary and sudden revolutions . . . of which we have no longer any examples."[17]

## The Contingency of the Past

The second formation of the earth was explained in a more precise fashion by John Woodward (1665–1722), the first author to suggest this concept. He proposed the idea of a dissolution of the earth before Bourguet in *Essay toward a Natural History of the Earth.* Woodward maintained that the cause of cohesion of bodies was, if not suspended, as his adversaries blamed him for saying, at least "diminished," which was when shells entered the rock formations. His explanation was simply that "gravity is entirely in the hands of God" and that he can suspend it at will.[18]

The theories we have examined, starting with Descartes', constrained God to a certain constancy (one might say here a certain coherence) in his manner of ruling the world. Woodward accepted that God might change or at least temporarily suspend certain laws he had given to nature. Thus, the past became *contingent* (dependent on chance or conditional on something uncertain) and we cannot reconstruct it based solely on the knowledge of natural laws and its present state, as Descartes invited us to do.

How can we reconstruct the past if it does not depend strictly upon natural laws? By looking into its archives, answers the historian who, without denying the existence of laws, knows that they are not sufficient to learn about past events. Geological archives were not useful to Descartes because he believed that the past, the present, and the future of the world could be understood by reason alone. However, if we are a little more modest in our ambition to penetrate the knowledge of natural laws, archives become useful to historians of the earth—even if they do not believe that God departed from the laws he had imposed on the world, and even if they are atheists.

In fact, Descartes was not interested in the *history* of the earth, but rather in its *formation*. However, the past cannot be deduced from the knowledge of simple laws because it is too complex. Consequently, the historian of the earth, as of human history, cannot totally neglect archives.

The archives of nature are fossils and the layers that include them. Descartes ignored them. However, even before Burnet and Woodward, a naturalist was able to recognize the importance of these archives: Nicolaus Steno.

# Chapter 5

# The Birth of Science

## Nicolaus Steno

In 1669 there appeared in Florence *De solido intra solidum naturaliter contento dissertationis prodromus* (The prodromus to a dissertation concerning a solid naturally contained within a solid).[1] Its author, Niels Stensen, or in Latin Nicolaus Stenonius (1638–1686), had left his native Denmark five years earlier to travel first in France and then in Italy. He lived for some time at the court of the grand duke of Tuscany, Ferdinand II, in Florence, where he renounced the Lutheran faith and converted to Catholicism. In 1669 he traveled to Austria and Hungary visiting mines with the intention of writing the treatise promised in his *Prodromus*. In 1670 he went to Holland and then returned to Florence. However, he did not write his main work. Considering the new ideas contained in the *Prodromus*—whose strange title is not self-explanatory to the modern reader—we can only regret his decision deeply.

Historians of biology consider Steno an anatomist, known particularly for his description of the duct of the parotid gland (named after him, Stensen's duct) and for his studies of muscles. However, geology owes much more to him.

Steno came to geology by way of fossils, and to fossils by the dissection of a large shark caught near Leghorn, Italy, in October 1666. Examining its mouth, he found teeth that resembled the so-called

---

# From Serpent Tongues to Shark Teeth

Shark teeth, known as *Glossopetrae*, have long intrigued people since Pliny reported that they were believed to have fallen from the sky on moonless nights. Some people saw in them human tongues, others saw serpent tongues, and still others saw bird tongues. The legend of serpent tongues supposedly started because large teeth were found on the island of Malta, where they "led to the belief that they were serpent tongues produced by sports of nature, whereas the smaller ones of these teeth were serpent eyes. The inhabitants of the island happily continued this legend when selling these fossils to strangers, saying that according to local traditions St. Paul cursed the serpents of the island when he was bitten by one of them. As attested by the Bible, the Earth being no longer able to produce serpents, made tongues and eyes of these animals instead, in memory of the miracle by the great apostle when he was bitten by a viper."*

---

*F.M.P. Colonna, *Histoire naturelle de l'univers*, 4 vols. (Paris: A. Cailleau, 1734), 2:301.

---

*Glossopetrae,* meaning petrified tongues, fossils so called because their shape recalls the tongue of a serpent.

The organic origin of fossils was far from being well established. In spite of the intuitive ideas of Leonardo da Vinci, mentioned earlier, as well as the slightly less bold ideas of Bernard Palissy, many contemporaries were not convinced that shells originated in the sea. Even increased observations did not remove the doubts of skeptics. Quite to the contrary, especially when forms were discovered that had no modern equivalent.

## Lost Species or Freaks of Nature

Today, evolutionary paleontology teaches that species evolve over time and that some are extinct and replaced by better adapted ones. It is therefore not surprising to find fossils that differ in form from present-day species. However, eighteenth-century naturalists were not ready to accept such an interpretation. Leonardo da Vinci had tried to convince his contemporaries—at least those who knew about his ideas—that fossils were remains of organisms. His interpretation was based on the identity of shells buried in rocks and those found on beaches. However, such proof was not valid when fossils and living organisms had different forms.

Bernard Palissy was not upset by this difference in form. Noticing "horns of Ammon" (**ammonites**, or fossil mollusks related to the *Nautilus* of the Indian Ocean), he stated that they had no present-day equivalent. He concluded that "their species was lost," probably because it was caught too often by fishermen.[2] One does not know whether to admire the perspicacity of this potter who saw fossils as a unity in the organic world, or to be annoyed that he solved with such candor the formidable problem of lost species. The words he used gained a certain fame: "lost species" were mentioned until the end of the eighteenth century.

It is understandable that not everybody was convinced by Bernard Palissy's argument, and even more so when it was found that fossils vary according to the rock unit in which they are buried. The easiest way to explain the specificity of fauna as a function of the formation in which they occur was naturally to think that fossils are *lapides sui generis*; that is, that they were formed as rocks and were never part of any animal whatsoever. Martin Lister, an English contemporary of Steno, supported this opinion with what he felt were solid arguments.[3]

Nevertheless, not all who refused to believe in the organic origin of fossils based their arguments on such careful observations. In the middle of religious wars, people searched in nature for signs of God's support for their respective faith. Indeed, some observers saw figures reminiscent of the papal tiara in fossils; others discovered Luther's face. Still others, with different preoccupations, found male or female sex organs.

Against these products of imagination, there is no other method than good observation. Steno studied the shark's head with great care, and in his report on the dissection in 1667, he tended toward belief—while pointing out that his interpretation was tentative—in the organic origin of *Glossopetrae*, which are so similar to fish teeth. He adopted an argument contrary to Lister's and stated that fossils found in rocks resemble each other no matter what kind of rock forms that layer, and that, furthermore, they are identical to "parts of animals which they resemble."

In 1669 Steno said in his *Prodromus*, now quite sure of himself, that he could attest that "those bodies which are dug from the earth and which are in every way like the parts of plants and animals, were produced in precisely the same manner and place as the parts of the plants and the animals were themselves produced."[4]

## Hooke and Leibniz

Steno's demonstration was important; it did not, however, stand alone in the second part of the seventeenth century. One year after Steno's *Prodromus*, the painter Agostino Scilla denounced as "vain speculations" the ideas of those who saw in fossils freaks of nature. Scilla's book, *La vana speculazione disingannata dal senso* (Vain speculation proven wrong by common sense), had much influence on his contemporaries because of its beautifully drawn fossils and clear language.[5] Two authors of higher status were less influential because their works were not known during their lifetime. They are Robert Hooke (1635–1703), an English physicist and naturalist who was an important member of the Royal Society, and the German philosopher Gottfried Wilhelm Leibniz (1646–1716), inventor of calculus at the same time as Newton. Hooke vigorously denounced the idea that fossils "are generally believed to be real Stones form'd into these Shapes, either by some plastick Vertue inherent in these Parts of the Earth, which is extravagant enough, or else by some Celestial Influence or Aspect of the Planets operating at a distance upon the yielding Matter of the Parts of the Earth, which is much more extravagant."[6]

Hooke went even further than Steno in regard to "species of organisms which have no present-day analog." Since these organisms seemed to support the thesis according to which they were formed by some plastic virtue that acted inside rocks, he explained more simply that these species had been destroyed. He even considered "that there may be divers new kinds now which have not been from the beginning."[7] This was certainly a bold statement, with respect not only to science but also to theology because it opposed the generally accepted idea of a single Creation.

Leibniz did not lack courage either. After having observed lost species, he formulated a daring hypothesis for his time: "A large number of forms were transformed during the great changes to which the Earth was subjected."[8] Without anachronistically identifying this concept with modern "transformism," we might say that evolutionism could find at least some seeds in this expression.

Nevertheless, these intuitions seem to have lagged way behind

Steno's trailblazing ideas. At any rate, Steno's concepts were of direct and immediate interest to geologists, whereas Leibniz's and Hooke's forecasts really concerned evolutionary biology or paleontology, subjects that became important only in the nineteenth century. Furthermore, a glimmer traced in a part of a sentence can hardly be compared to a precise argument as developed by Steno in a paper of sixty pages. Finally, even if some parts of these naturalists' contributions were identical, Steno's work had a more immediate success. Indeed, Hooke's ideas on earthquakes and volcanoes were only published posthumously in 1705; the *Protogaea*, of which Leibniz had merely given a two-page abstract in 1693, was not expanded and published in full before 1749, thirty-three years after the author's death. It appeared at the same time as Buffon's "Théorie de la terre" (included in *Histoire naturelle, générale et particulière*, Natural history, general and particular), when the problems about the origin of fossils had changed to a certain extent (although the French naturalist, as shown in chapter 7, still felt compelled to prove that fossils were remains of organisms).

## Principles of Geology

Let us return to Steno; otherwise the reader shall begin to lose patience in waiting for the great discovery I announced at the beginning of this chapter. The discovery is in fact a simple one; namely, to draw the consequences from the origin of fossils. Steno reasoned that, if fossils and other fragments are the remains of ancient living beings, the rock unit that contains them must have been deposited on the seafloor. Leonardo da Vinci had understood that much. However, Steno went further and distinguished two kinds of environments of deposition: "If in a certain strata we discover traces of salt of the sea, the remains of marine animals, the timbers of ships, and a substance similar to the bottom of the sea, it is certain that the sea was at one time in that place." If, on the contrary, "we find a great abundance of rush, grass, pine cones, trunks and branches of trees, and similar objects, we may rightly conclude that this matter was swept there by the flooding of a river or the inflowing of a torrent."[9] We might smile at the description of timbers of ships being treated on the same level as fossils dating back tens or hundreds of million years, as if human endeavors had the same age as the earth. However, Steno's universe was that of the Bible: it was only some thousands of years old. Nevertheless, he built into that short time span the bases for nineteenth-century geology because he assigned to fossils the role of defining the **facies** of geological **formations.**

His boldness did not stop there. He said that, because fossiliferous layers are aqueous deposits, their surfaces must be horizontal. Even when the substratum was irregular, the first deposit would level it, and its surface would be horizontal. Therefore, "all strata, except the lowest, are bounded by two planes parallel to the horizon."[10] Furthermore, he maintained that successive layers are increasingly younger, going upward in the series. This law, called today the principle of superposition, states that the age of layers decreases from bottom to top. Although this principle appears self-evident to any student of geology, the fact remains that nobody had thought of stating it before this extraordinary man.

The second principle, original horizontality, is just as simple. Whereas the first one concerns stratigraphy (order of deposition of strata), this one concerns tectonics (deformation of strata after deposition). Steno stated that "strata either perpendicular to the horizon or inclined toward it, were at one time parallel to the horizon."[11] In other words, sedimentary formations are originally deposited flat. Any dip they now have has resulted from subsequent folding or tilting.

Readers who recall their first lessons in geology will surely remember studying these laws. In these truly modern laws, Steno laid the groundwork for the method of investigation of layers (or strata) of the earth, the fundamentals on how to read the archives of nature.

## Tilting of Beds

For a good understanding of this new discovery, we should return to Descartes. He explained mountain building by collapse, where the slopes of mountains represented the effects of that collapse. This theory was quite original but purely hypothetical. Steno was not content to make conjectures, and he invited the reader to observe not the slopes of mountains but the tilting of beds in these mountains. From this fact, he deduced the movement that placed them in that position.

Nevertheless, Steno agreed with Descartes on one point, which we now know to be erroneous: the direction of movement. Like Descartes, he believed that tilted strata had been caused by the collapse at one end of the layers rather than by uplifting of the opposite end;

we know today that mountains are being formed as a whole by an uplifting movement.

In fact, at that time it was difficult to imagine uplifting of beds. True, lavas did erupt from great depths during volcanic activity, but this upward movement was explosive and often compared to the combustion of gunpowder. Its effect would be the rupturing rather than the uplifting of layers. It could therefore not be responsible for their tilting.

Tilting was more easily explained by the removal of underlying layers that had supported the upper layers. Upon withdrawal of their foundation, the upper strata were forced to collapse. Like Descartes, Steno explained the formation of plains but not of mountains—but topographic relief is a relative matter. What caused the undermining of the foundation of sedimentary rocks? Steno believed in the action of fire or water without giving further details.

## The History of Tuscany

To appreciate the superiority of Steno over his predecessors, just read, in the last pages of the *Prodromus*, how he applied his principles to "changes which have occurred in Tuscany," and the illustration of these changes in six figures. It is true that Descartes also illustrated his explanation with plates. However, they were on a global scale, without reference to any real topography. Burnet also included diagrams, but with the exception of the replacement of the central fire by earth, the result was the same as Descartes'.

Although book illustrations were rather scarce in the seventeenth century, there were some exceptions in the form of completely abstract diagrams of the globe. For instance, Athanasius Kircher (1602–1680) illustrated with several plates his *Mundus subterraneus*, which appeared shortly before the *Prodromus*.[12] Although he remained geocentrist thirty years after Galileo's death, this Jesuit priest provided the globe with a central fire spewing out lava at the surface through volcanic craters representing many outlets for this interior force.

Steno did not show the whole globe but limited his representation and commentary to Tuscany. Moreover, although Descartes described the formation of the earth in a direct sense, starting from a postulated beginning, Steno presented his diagrams in a retrograde sense, going backward from the present state to former ones. This

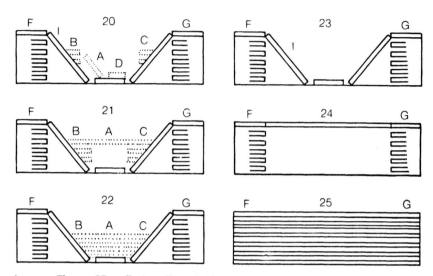

LEGEND: Figure 25 = first sedimentation (stony layers, in solid lines); Figure 24 = emergence and undermining; Figure 23 = collapse of the upper layer; Figure 22 = new sedimentation (sandy layers, in dotted lines); Figure 21 = emergence and undermining; Figure 20 = collapse of the upper layer.

Figure 5.1. Various Changes That Occurred in Tuscany (From Steno, De solido intra solidum . . . , 1669, plate II). In B and C, horizontal sandy layers rest on tilted rocky layers (angular unconformity).

inductive reasoning assumes observation of present-day structure. Unfortunately, we don't know what Steno understood about this structure, although he said that he recognized "six ancient aspects of Tuscany . . . by inference from many places examined by me, so do I affirm it with reference to the entire earth, from the descriptions of different places contributed by different writers."[13] His figure 20 is too abstract to give any useful information. This is why it is perhaps simpler to describe his six figures in a reverse order; that is, to give the history of Tuscany in a direct sense (fig. 5.1).

Indeed, the area went through six successive stages: two immersions, two emergences resulting in a flat and dry surface, and two stages during which Tuscany acquired uneven landforms. Together, these stages represent two geological cycles (cycle of sedimentation and cycle of uplifting). The examination of these six figures, starting with his last one (i.e., the first stage), shows what follows (see fig. 5.1):

1. The area is submerged and a thick sedimentary series is deposited (figure 25 of original plate)

2. The area has emerged and its lower layers are undermined (figure 24)
3. Collapse of the upper deposits (figure 23)
4. The sea returns and new sedimentary strata are deposited in the basin formed by the first collapse (figure 22)
5. The sea retreats again and internal sapping of the new deposits occurs (figure 21)
6. Finally, a second collapse (figure 20) intermingles younger layers with the older ones that collapsed during the first cycle

## Defining Cycles

Based on the present structure of Tuscany, we may ask how it is possible to recognize the two geological cycles because their tilted layers are today intermingled. Steno distinguished them first by their nature: deposits of the first cycle were formed by rocky layers, whereas those of the second were sandy layers. This is shown on the figures by solid lines for the first cycle and dotted ones for the second.

Nevertheless, we may question again why Steno did not use his principle of superposition. His figure 22 shows clearly, for instance, that sandy deposits were at the same height as earlier deposits. The principle is therefore valid only within a cycle. But how did he distinguish the two cycles?

Looking closely at Steno's figure 20, we see a fragment of a sandy layer resting horizontally, at point B, against an inclined rocky layer. This might evoke what we call today an **angular unconformity** (chapter 9). However, Steno did not seem to pay any particular attention to this observation. We must regret once more that the work announced in the *Prodromus* was never written, because it might have answered some questions about the nature of Steno's investigation in the field and the extent of his perspicacity in the application of the principles he so clearly expressed.

Steno showed himself as innovator in yet another field. As mentioned earlier, he guessed that fossils were of interest in determining whether ancient deposits were of marine or terrestrial origin. It is known today that fossils are the keys to dating layers (see stratigraphic fossils, chapter 10). Without assuming that Steno used fossils to recognize the ages of rock units, we should at least notice his statement that the oldest layers are free from traces of life, which is indeed a step in that direction.

However, he surprises us when he says that layers "free from all heterogeneous bodies" are found "in the highest mountains."[14] We might expect that he would rather place them in the bowels of the earth because these layers, deposited at a "time when neither plants nor animals existed," were the first formed. Steno did not use his principle of superposition in this case either.

Steno's structural interpretation, calling for stepped successive cycles, was very conjectural and he himself did not elaborate upon it. Furthermore, in the framework of his time, he could hardly do so: such an explanation would not be consistent with the author's clear statement about his loyalty to Scripture. Indeed, from the beginning he stated, "Fearing that one would be afraid of seeing things in such a novel fashion," he was going to show the agreement between nature and Scripture. Therefore, his notion of time was that of any pious reader of the Bible in his day. The shells he found in the Etruscan city of Volterra "were three thousand years old or a little more," and with them "we can easily go back to the very times of the universal deluge."[15] Steno, like his contemporaries, knew how to be content with short time spans, which left little time for repeated inundations.

## The Archives of the Earth

Despite all his limitations, Steno provided a considerable leap forward in methodology for histories of the earth. The fact that some rock layers were tilted or otherwise deformed called no longer for speculation, but for observation. Naturalists could no longer simply produce conjectures in their studies, as Descartes had done on the formation of the earth. They had to go into the field to observe and collect proofs of the past from the "archives of nature." And they did: Peter Simon Pallas (1741–1811), who explored mountains in Russia, in particular the Urals and the Altai, between 1768 and 1774; and Faujas de Saint-Fond, who explored mountains in France at about the same time. Although Steno himself did not use the term "archives of nature," what he said corresponded fairly well to the way the phrase was used a century after him. Buffon, in his "Époques de la nature," spoke of the "archives of the world"; and J. L. Giraud Soulavie (1752–1813), one of the founders of paleontologic stratigraphy, talked about the "annals of the physical world."[16]

These metaphors were certainly appropriate, since archives and annals are both historical documents. Steno applied to the history of the earth what historians did for human history. At last, theories on the formation of the earth were replaced by studies of its history.

Similar expressions became household words for geologists only a century after the *Prodromus*. It took that long after Steno's death for naturalists to realize the importance of his achievement—evidence that he was not understood during his century.

Some historians of geology have noticed that Robert Hooke, a contemporary of Steno, used a similar metaphor when he said, "Now these shells and other Bodies are the Metals, Urnes or Monuments of Nature."[17] I do not deny the important role played by Hooke, already mentioned at the beginning of this chapter; but unlike Steno, Hooke made no use of the archives. He had perhaps an intuitive sense of the value of fossils in stratigraphy because, as we have seen, he suspected that old species might disappear and (above all) new ones appear. However, it was premature to apply the idea of new species: fossils and living organisms could hardly be compared as yet because so few fossils had been studied carefully. Steno's principle of superposition was infinitely more fruitful because immediately applicable. If genius is measured by the use made of new ideas, in geology Steno was far superior to Hooke even though his discovery was not applied by his immediate successors.

Steno thus paved the way for the writing of histories of the earth; that is, the reconstruction of the past based on proofs, or "monuments" as they were often called in the eighteenth century. These proofs required a certain contingency of history. Indeed, most geological events cannot be forecast because they do not occur according to general laws. Therefore, one cannot do without "monuments," or proofs. And this is the work of a historian. A. Cournot has made an excellent distinction between historical investigation and studies based on certain laws and facts: "The description of phenomena whose stages necessarily follow each other and are linked together according to laws pertaining to reasoning or experience belongs to the domain of science."[18] On the contrary, "many events have occurred in the past which according to their nature cannot be investigated by a theory based on facts or on the knowledge of permanent laws. They can be known only by historical investigation."[19]

After Steno, the earth's past entered into the category of facts that could be investigated historically. The classification of mountains into two groups of different age was to be the first step in that direction.

# Chapter 6

# On Mountain Building

## Exploring the Mountains

In the eighteenth century, particularly during the second half, nature was discovered by the public at large. Jean-Jacques Rousseau not only set the fashion for gathering plants, but in his novel, *La Nouvelle Héloïse* (1761), he also led his readers to the discovery of mountains. Indeed, in his description of the Valais in Switzerland, he evoked alpine landscapes that appeared stranger to many of his contemporaries than do those of the moon to us: "Now huge rocks were hanging in ruins above my head, now I was submerged by the thick mist of high and roaring waterfalls, now an endless torrent disappeared into an abyss close to me, the depths of which no eye would dare to explore."[1]

Whereas Rousseau traveled "in ecstasy through these little known places which are so worthy of admiration," Horace-Bénédict de Saussure, at the age of twenty, promoted a daring project: the ascent of the highest European summit, the dangerous Mont Blanc, which towers above the valleys of Chamonix and Courmayeur. He accomplished the task only a quarter of a century later.

In August 1787, at the head of a field party, he set foot on the snow-covered summit. Jacques Balmat, a guide from Chamonix, and the young doctor Paccard had reached the much desired peak a year earlier, but to athletic achievement Saussure added scientific observations (see box).

---

## The Climbing of Mont Blanc
## by H.-B. de Saussure

Finally, the moment I had wished for arrived when I started out on August 1 [1787], accompanied by a servant and eighteen guides who carried instruments of physics and all the paraphernalia I needed. My older son had wished very much to come along; however, I was afraid that he was not yet robust enough. . . . He stayed at the priory, where he carried out careful observations similar to those which I made on the summit. . . .

We reached at a quarter till two the summit of the mountain La Côte, the place where we were to spend the night. The first day had thus not been long. It had taken us only six and a half hours from the priory to our first camp. . . .

The following day, August 2, in spite of our eagerness to leave very early in the morning, so many disputes arose between the guides about the distribution and the arrangement of their loads that we were actually off only at six thirty. . . . We began to cross the glacier, opposite the granite boulders where we had taken shelter for the night. . . . It took us almost three hours to cross this dangerous glacier, although it is hardly a quarter of a league wide. . . .

After a one-hour walk, we had to coast along a huge crevasse. It was over a hundred feet wide, and we could not see the bottom. [After having crossed the crevasse, then dined and gotten over a first plateau,] we climbed for almost an hour on a 34° slope and finally reached the second plateau, where we planned to spend the night. . . .

---

At the same time, L. F. Ramond de Carbonnières (1755–1827) reached Mont Perdu, a summit of the Pyrenees. As with Saussure, his ambitions were above all scientific. Mountains now became an attraction for meteorologists (Saussure carried with him barometers and hygrometers) and geologists. They now considered mountains worthy of detailed studies.

As mentioned above, Steno placed rock layers without "heterogeneous bodies" at high altitudes. Although today this statement may seem out of place in the writings of the founder of the principles of stratigraphy and tectonics, it soon struck neptunists as logical.

## Telliamed

Benoît de Maillet (1656–1738), French consul at Cairo, has won a place in the history of natural sciences through an anonymous

*(continued)*

The following day we first crossed the second plateau . . . from there we climbed to the third. . . . After a walk of two and a half hours, we reached the boulder that I call the left shoulder or the second staircase of Mont Blanc. There I could see an immense horizon completely new to me. . . .

From Chamonix I had measured the elevations of the different parts of the mountain, and hence knew that only about 150 toises were left for me to climb. . . . I hoped therefore to reach the summit in less than three quarters of an hour; however, the thin air gave me greater difficulties than I had anticipated. Finally, I was forced to catch my breath every fifteen or sixteen steps. . . . At last I reached the goal I had been striving for so long. But during the two hours of this difficult climb I had already looked at almost everything that I finally saw from the top, so that the arrival was no longer a sensational event.

Nevertheless, the view of the mountains gave me a vivid satisfaction. . . . It seemed like a dream when I saw these needles of Midi, of Argentière, of Géant, the basal slopes of which had been for me so difficult and so dangerous to reach.*

---

*Horace-Bénédict de Saussure, *Voyages dans les Alpes, précédés d'un essai sur l'histoire naturelle des environs de Genève*, 4 vols. (vol. 1, Neuchâtel: Samuel Fauche, 1779; vol. 2, Geneva: Barde, Manget & Cie, 1786; vols. 3 and 4, Neuchâtel: Louis Fauche-Borel, 1796), vol. 4, § 142, 147, 155–157, 160, 163, 168, 171, 175–176.

manuscript he put into circulation in the 1720s under the title "Nouveau système du monde ou entretien avec Telliamed" (New system of the world or discourses with Telliamed). Telliamed is his name spelled backward. The work was only published in 1748, ten years after his death.[2] This diplomat's thesis was quite daring: he believed that humans had descended from marine ancestors. As proof, he offered travelers' tales of sea monsters with human traits. Some historians of transformism have considered this author a forerunner of Darwin, although the legends and stories given as proofs of Maillet's extravagant theory seem a little far-fetched and hardly worthy of being taken seriously. However, his geology was a great leap forward and places his views clearly at the beginning of the distinction between two successive kinds of mountains.

His main idea is this. The sea level keeps getting lower; therefore, all terrestrial species descended from marine ancestors that underwent transformation during the diminution of the sea. His great audacity lies in measuring time by thousands of centuries and in believing that the diminution of the sea lasted 500,000 years. (A contemporary of Maillet, under the pseudonym of the "Turkish spy," a name assumed by various authors, also reckoned enormous time spans.) However, of interest to the geologist is Maillet's use of that time. As the sea diminished, mountains emerged that were increasingly younger and less elevated.

The oldest mountains, which he called "primitive," contained no fossils. However, Maillet's reasons for adopting the Stenonian concept were different from Steno's. Maillet did not believe in a creation of the world. According to him, the earth had undergone previous cycles of drying up and rehydration; mountains were primitive only in the sense that they were old compared to the present world. Nevertheless, marine organisms could live only in shallow shoreline water. As a result, when the sea level was much higher, these mountains were too deep under water to allow any sort of life.

How were these mountains formed? By marine currents at the bottom of the ocean. The materials accumulated there derived from the reworking of the ashes from an earlier extinguished sun whose indurated crust formed the ocean floor. During the shrinking of the waters, the highest parts of the mountains emerged and life appeared in these areas. New deposits were formed from the remains of living organisms, and hence new mountains were born. They leaned against the older ones and were not as high.

The evolution of the earth's crust according to Maillet's system has been interpreted by A. V. Carozzi (see figs. 6.1 and 6.2).[3] The diagrams show successively younger mountains leaning against older ones; that is, a decrease in elevation from the oldest to the youngest mountains. The modern reader who has learned that young mountains are generally high because they are not eroded whereas old mountains are lower because they were peneplained through time may be surprised by this arrangement. Nonetheless, it is worth noting that Maillet's interpretation does explain some mountain structures better than the usual modern explanation of the topographic relationship between old and young mountains.

## A Cross-section of the Alps

Up to this point, all we had to do to follow the debate between early geologists was accept that certain rocks are fossiliferous. From now

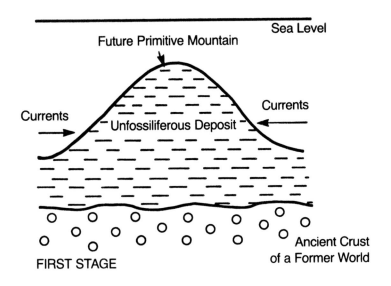

FIRST STAGE

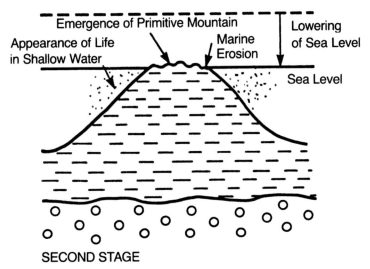

SECOND STAGE

Figure 6.1. Stages 1 and 2 of the Formation of Mountains According to the Concept of B. de Maillet (Modified from Telliamed, 1748, trans. A. V. Carozzi, 1968, reproduced by permission of the University of Illinois Press, Urbana).

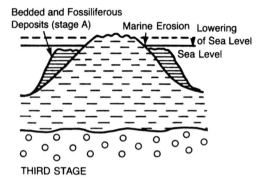

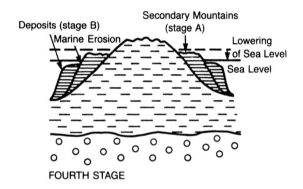

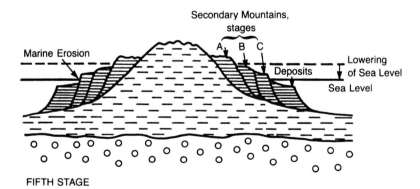

Figure 6.2. Stages 3 through 5 of the Formation of Mountains According to the Concept of B. de Maillet (Modified from Telliamed, 1748 trans. A. V. Carozzi, 1968, reproduced by permission of the University of Illinois Press, Urbana). The five diagrams correspond to five stages of successive lowering of sea level.

on, more knowledge is needed. Let us pretend to go westward from Mont Blanc along a straight line. We would encounter a few other summits higher than 4,000 meters before crossing the High Calcareous Alps, which are at a lower elevation, and reaching the peri-Alpine plains and plateaus. Mont Blanc and the adjacent summits consist of igneous and metamorphic rocks (granite, gneiss, and others), which are neither fossiliferous nor stratified. The High Calcareous Alps, on the other hand, show folded fossiliferous layers. This situation is more or less as Maillet describes it, and occurs all along the Alpine chain.

Maillet's interpretation fails us, however, when we consider the origin of these two types of mountains. Maillet believed that the older mountains, those consisting of igneous and metamorphic rocks, such as Mont Blanc, acquired their present height during their original deposition. According to modern theories though, these mountains underwent two successive orogenic uplifts. One occurred near the end of the **Paleozoic era,** some 250 million years ago when all of middle Europe was transformed into a vast mountain chain called **Hercynian mountains.** The present topography is not a residue of these huge Hercynian mountains because these were peneplained at the end of the Paleozoic era, about 225 million years ago. Only their substratum remained, and new oceans poured over these lands. Erosion destroyed the upper parts, and today this substratum, or basement, consists only of the deeper parts of the Hercynian mountains, those transformed into igneous and metamorphic rocks by heat and pressure.

The second orogeny began toward the end of the **Mesozoic era** when the circum-mediterranean chains were formed, stretching from the Pyrenees to the Taurus and including the Alps, the Carpathians, the Atlas chain, and others. Mountain building continued during part of the 60 million years of the **Cenozoic era.** This uplifting exhumed the cores of the old Hercynian chain; from these cores the "cover" of younger layers (High Calcareous Alps, for instance) slid down as gravity thrusts (**gravitational sliding**).

When I said earlier that old mountains are often lower than new chains, I was actually not referring to the Alps but to such old mountains as the Massif Central or the Vosges in France or the Appalachian Mountains in the United States, which owe their present relief to movements more or less contemporaneous with the Alpine orogeny. These mountains—as well as the Alps—were eroded and

---

## Basement and Sedimentary Cover

When sedimentary layers rest **unconformably** on folded meta-morphic rocks that were injected by igneous rocks during a former orogeny and thereafter peneplained by erosion, the upper layers are called sedimentary cover and the lower ones basement. The Central Massif and the Armorican Massif thus form the exposed parts of the basement of the Paris Basin.

When an area undergoes tectonic stresses, the previously highly folded and strongly crystallized basement of igneous and metamor-phic rocks breaks up in a brittle fashion, whereas the more ductile sedimentary cover is deformed. If stresses are powerful and if layers conducive to sliding exist at the base of the cover (for instance, very ductile clays), the cover separates from the basement and undergoes deformations completely independent from those of the basement. For example, the High Calcareous Alps represent the sedimentary cover of the igneous and metamorphic rocks of Mont Blanc, which slid away northwestward as gravity thrusts, leaving behind the uplifted basement.

---

peneplained at the end of the Paleozoic era and uplifted only re-cently, but of course not as much as the Alps.

## The Wonders of High Mountains

The interest in high elevations did not always have a rational basis. At the end of the eighteenth century, many people were forever praising mountains for sentimental or aesthetic reasons. J.-A. Deluc called them "a wonderland" where "all is beautiful and lavish."[4] Dé-odat de Dolomieu, even more enthusiastic, did not hesitate to de-scribe all mountain folk as "good fathers, good sons, good husbands, and good parents."[5] He showed quite incidentally the ambivalence that mountains could engender in the human spirit. "Now," he said, "you might believe that you are attending the creation of the world, now you seem to look at its ruins."[6] Youth and old age seemed to meet in this strange and unfriendly world (all travelers suffered from high altitude), which linked the purity of childhood to the decrepi-tude of old age.

If "reason" invited naturalists to search in mountains for traces of the first ages of the history of the world, they often took as their start-ing point the view that the earth was initially covered by a chaotic ocean, "formed," as the Bible says, "by means of water." Although Maillet was far removed from religious preoccupations, his neptunist theory of the origin of mountains still bore the imprint of this cultural theme.

He believed also, as did many others, that old mountains were of high elevation. In the case of Mont Blanc this statement still holds; it is not paradoxical to climb the Alpine summits in order to find the oldest deposits of the history of the earth. But the old granitic summits of the Alps reached their present-day location through uplifting, a concept Maillet never used.

All through the eighteenth century, mountains were generally believed to be simply superposed layers formed at the bottom of the ocean—hence the name of neptunism given to this concept. Primitive mountains were considered to be the summits of uneven deposits, or, in other words, the ancient layers taken as a whole formed primitive mountains. In 1752 the geographer P. Buache wrote an essay "On the kind of structure of the globe, composed of mountain chains which cross the oceans as well as the lands," a title reminiscent of the concept of the earth's bone structure used in earlier centuries.[7] The term *mountain* was often taken in a large sense to describe not only the highest peaks but also the entire deposit of the same age. Closer to the neptunist concept, the German language used the same word, *Gebirge* (or *Gebürge*), for both the mountain and the subterranean rock layers. In reality, primitive mountains described by neptunist authors referred to what is called today the Hercynian basement. Although their interpretation was erroneous, they did correctly observe the distinction between primitive mountains and the sedimentary cover.

The weakness in neptunism was the need to assume that summits of 4,000 meters or more were once beneath sea level, a situation requiring huge oceanic masses. For Maillet, who measured time by thousands of centuries, this water needed merely to dry up. However, authors who followed the Scriptures faithfully were burdened by the problem of having only thousands of years to accomplish this.

## Volcanic Mountains

In an attempt to overcome the problem of lowering sea level, the Italian Lazzaro Moro (1687–1764), contemporary of Maillet, stated that the primitive ocean did not rise more than 175 *toises* (about 350 meters) above its present level. He proposed that any mountains with strata and fossils higher than this elevation had been uplifted.[8]

He, too, distinguished two classes of mountains. The first ones "originated in the bowels of the Earth when the surface area where they were born was still covered by water and not burdened by terrestrial materials." In other words, erosion had not begun because no island had emerged. These mountains, which Moro called "primary," were formed by "large masses of rocks" and were thus not stratified. After their emergence, erosion produced materials that were deposited at the bottom of the ocean and were uplifted in turn to form "secondary" mountains. According to Moro, they differed from the first ones in their structure, which consisted "almost entirely of superposed layers."[9]

To explain uplifting, Moro referred to recent volcanic eruptions. He mentioned that an island emerged in 1707 and reminded his fellow Italians of the birth of Monte Nuovo in 1533, in the Bay of Naples, in the middle of the Phlegraean Fields, very close to Vesuvius. He added, since "the same causes produce the same effects," what happens today must have happened during the first periods of the history of the world.[10]

Moro thus spoke like a uniformitarian, and his theory seemed to offer a way out of the neptunists' dilemma. We might therefore expect that it would convince his contemporaries. Nothing of the sort happened. His attempt as a vulcanist reaped no success, and neptunism continued to rule among naturalists. It was only in the nineteenth century, when Charles Lyell raised the uniformitarian approach to an important concept (chapter 11), that Moro's ideas were belatedly appreciated.

## The Retreat of the Sea

We must admit that the neptunistic view also offered good arguments that could claim to rest on observed facts. In 1724 the physicist Anders Celsius (1701–1744), known for his invention of the Celsius scale, measured the level of the Baltic Sea. He placed benchmarks to measure variations of sea level.[11] One of the measurements made in 1731 by Celsius was used in 1747 to show that sea level had lowered in the Gulf of Bothnia.

The famous naturalist Carl Linnaeus adopted Celsius's thesis and published in 1744 an essay called *The Growth of the Earth*, in which he explained that "the entire *land* was during *the Earth's youth drowned in water* and covered by a vast ocean *with the exception of one island.*"[12] His goal was essentially biological: he placed the island at the equator, without great concern for geographical veracity, where it was to receive a pair from each species or one individual from hermaphrodites. However, and this is the interesting point, the famous botanist thought he had to calculate the retreat of the sea. He said that it had fallen 5 to 6 feet over a hundred years so that the sea level had dropped 240 to 300 feet in six thousand years. However, Linnaeus made a mistake of the order of ten in his calculations, writing that the sea had retreated 2,700 feet. Strange slip of the pen. He corrected it later to 240 feet, which reduced the mountain to an anthill.[13] There is no better way to demonstrate the inconveniences of traditional chronology.

These examples of eighteenth-century geological theories demonstrate how difficult it was to keep geological phenomena in the framework of biblical chronologies and to give them, at the same time, the magnitude required by observations of natural phenomena. The calculation error of Linnaeus reveals the obstacle he and his contemporaries encountered. Maillet overcame it easily by breaking away from short chronologies. Moro solved the problem in another fashion by abandoning neptunism in favor of vulcanism. Nevertheless, these hypotheses remained isolated and unused because both required a definite break with generally accepted ideas. Most naturalists of that period continued to associate biblical chronologies and neptunism, thus refusing both immense durations and uplifting movements of the earth's layers.

## Classification of Mountains

The mid-eighteenth-century author who contributed most to the establishment of the theory of two (or three) classes of mountains was the German mineralogist Johann Gottlob Lehmann (1619–1767). His book, *Versuch einer Geschichte von Flötz-Gebürgen* (Essay on a history of secondary mountains), published in 1756, was translated into French three years later.[14]

Lehmann said that in primitive mountains, which he called *Gang-Gebürge*, "strata are not horizontal but either perpendicular or diagonal. Beds are not as thin and varied as those in secondary mountains" (*Flötz-Gebürge*).[15] If any rare fossil shells occur in primitive mountains, they rest on the surface and are not incorporated inside the rock. Finally, these mountains are rich in mineral deposits, they are of high elevation, and they have steep slopes. Secondary mountains, or stratified mountains, are formed by horizontal, thin, and regular layers; they contain fossils buried in their rocks; they are poor in minerals, of low elevation, and their slopes are gentle. A third class of mountains exists, more insignificant yet, which Lehmann hardly described.[16]

Lehmann thus came closer to Maillet's than to Moro's ideas with respect to the origin of mountains. He believed that all mountains were formed in the oceanic waters and remained in the same location where they were deposited. The only difference from Maillet is Lehmann's belief that the two main classes were formed during successive advances of the sea.

In France, these ideas spread quickly thanks to the rapid translation of the Versuch[17] and the lively personality of the translator, who was no less than the baron d'Holbach, friend of Diderot and anonymous author of audacious philosophical works, such as Système de la nature, in which he defended atheism.[18] Interested in geology, d'Holbach was in charge of writing articles concerning the earth sciences in the Encyclopédie.

Guillaume-François Rouelle (1703–1767), friend of d'Holbach, supported similar ideas. In his lectures at the Jardin des plantes, he added a little bit of mineralogy and geology to chemistry and distinguished an "ancient Earth" with tilted layers, "which has always existed the way it is now," from a "new Earth," which was deposited on top.[19]

The same concept was used again by Giovanni Arduino (1714–1795), a mineralogist and chemist who distinguished primitive schistose mountains from secondary limestone mountains, and the latter from more recent ones, consisting of sand and clay, which he called tertiary mountains.

For all these authors, two characteristics were associated: decrease in elevation and change of the nature of rock formations from one class of mountain to the other. The nature of rocks and their location posed some problems. Lehmann identified primitive mountains by their veins and poor stratification. Undoubtedly, he meant granitic mountains, which are not stratified but in which some excellent observers believed they could see more or less regular layers (perhaps from seeing almost parallel fissures). Arduino did not find any granite in his primitive mountains in the Vicentino, but, knowing that they existed elsewhere, he wondered whether granite might not be underneath schists. In secondary mountains, limestone seemed to be the main constituent.[20]

The concept established by the three authors, each one for his own country, called for generalization. Because the second half of the eighteenth century saw the exploration of large mountain chains, the time was ripe to test in the field the model that worked for the border of the old Hercynian massifs in Central Europe. The Alps showed a rather similar arrangement and, as mentioned above, massifs of igneous and metamorphic rocks topographically dominate limestone massifs. Indeed, Saussure had no problem finding the arrangement of granite, schistose rocks, and limestone.[21] Only the Pyrenean chain seemed to present an anomaly.

## The Case of the Pyrenees

Pierre Bernard Palassou (1745–1830), who explored the Pyrenees in the early 1780s, found at first the same general arrangement.[22] However, in 1782 Déodat de Dolomieu (1750–1801) maintained that the center of the chain was limestone and not the expected granite. He was pleased because he opposed "systems" and deplored that one was "wanted to subject nature's productions to a certain order."[23] He was, by the way, an original thinker who held a notable place in geology at the end of the century. Without any doubt, he would have left an important work had he not dispersed his energy and become mixed up with various, sometimes obscure adventures. A serious quarrel with the Order of the Knights of Malta, to which he belonged, kept him in prison for twenty-one months beginning in 1799. He left prison very weak and sick and died in 1801 at the age of fifty-one, just when he had started to take up geological observations again.

However, Dolomieu had based his ideas on only one brief visit to the Pyrenees. It was one of his friends, Philippe Isidore Picot de Lapeyrouse (1744–1818), who confirmed Dolomieu's ideas in the last years of the eighteenth century after having observed that "primitive limestone," a "contemporary of granite," did "not contain the least trace of organic beings"—the undeniable proof of its old age.[24]

Another Pyrenean geologist, L. F. Ramond de Carbonnières, refuted this opinion. According to him, limestone was not a primitive but a secondary rock formation. Therefore, it must be fossiliferous everywhere. He challenged Picot de Lapeyrouse by proposing the two climb Mont Perdu, the main summit of the limestone chain, together.

The trip got underway, and Ramond won twice. Not only did he find the predicted fossils, but his companion could not follow him to the top of these unfriendly mountains. This was the year 1797, ten years after Saussure had reached the summit of Mont Blanc. Ramond was able to shout in triumph: "There are marine bodies on the ridge of the Pyrenees and on the peak which dominates all Pyrenean mountains." [25]

As a result, limestone mountains were no longer regarded as primitive mountains. They were younger than the granitic axis of the chain and the two adjacent schistose chains at its flanks. The Pyrenees thus became part of the generally accepted arrangement. The only anomaly they presented was the elevation of the parallel ridges: the northern adjacent schistose chain is highest at Vignemale, right above the axis, whereas the Vignemale itself is dominated by Mont Perdu. The order was preserved though the hierarchy between mountains was slightly amiss. Ramond attributed this abnormal altitude to an accidental accumulation, and the Pyrenean chain joined the geologists' list of mountains formed according to the common rule.

## Universal Order?

Before closing this chapter, a major problem has to be emphasized. It was generally accepted that, if the Pyrenees indeed showed the same order in the arrangement of layers as the Alps—granite, schists, and limestone—and if exploration of other mountains throughout the world showed that they belong to the same class, then the arrangement of strata had to be identical everywhere on earth (Pallas examined the Urals in the 1770s, and Humboldt crossed the Atlantic a little later to study mountain chains in South America).[26] The order of superposition of rock formations was therefore believed to be universal. As a result, it no longer seemed necessary to establish this order in remote and unfriendly mountains. Generalization was believed possible even when it was based merely on data already collected so long as that data seemed significant enough to prove the universality of succession. New observations would only confirm what was already probable.

If that were the case, did the layers of the earth still remain archives? The role of archives is to provide information about events that can be known only through them. However, if the order were universal, documents found in a limited area could give information about the history of the entire earth.

Neptunists were thus edging toward a system where archives would again become useless. And this in spite of Maillet, who had stressed the importance of geological relicts of the past. Maillet's view seems to have been neglected by historians of geology, and it is in fact rather unexpected coming from an author of a theory of cycles who was more interested in finding regular and repetitive events than random and unique ones. After noticing that in general the earth perished in flames at the end of each cycle, he wrote: "If we could dig down to the center of our globe, and go through the various arrangements of the materials forming it, we should be able to judge by means of these investigations whether it has been several times completely covered with water after having been inhabited, without having been the prey of flames. In such a case, one should find inside the globe all the remains of several superposed worlds." [27]

In other words, if the cycles were incomplete instead of following each other in an identical fashion, they would add their remains in a stratigraphic column. This theory foreshadowed the concept of sedimentary cycles used at the end of the century. However, Maillet stopped there. His theory as a whole did not lead him to pay attention to this progressive construction of the earth. Such a theory would have to stress the irreversible nature of the earth's history.

# Buffon as Historian

## The Hierarchy of Mountains

Lehmann's three kinds of mountains were of decreasing size. The structure of primitive mountains therefore ruled over everything else and provided the main feature of the earth's topography, to which secondary and tertiary mountains merely added slight alterations. But this was really not a new concept; Steno had been of the same opinion, although his system was very different.

Lehmann had no difficulties in explaining that hierarchy. Primitive mountains, he said, result from Creation, secondary mountains from the Deluge, whereas the last class was caused by upheavals, such as earthquakes, volcanoes, or floods, that still occur today. He believed that the earth owed its structure to the direct intervention of God, who had twice covered the earth with the waters of the ocean.

In the century of the Enlightenment, this unscientific concept could hardly receive unanimous approval. Rouelle refused to invoke the Deluge to explain the formation of the earth. He preferred to search for some sort of causes that acted daily at a slow pace, that is, present-day causes. He was a uniformitarian like his friend d'Holbach, whom we would easily suspect of not believing in the biblical Deluge.

Nevertheless, the mountains according to Rouelle and d'Holbach were just as hierarchic—that is, arranged by order of decreasing size—as those of Lehmann, and so were those of another freethinker: Maillet. Apparently it was not easy to abandon traditional ideas. Moreover, ideas of hierarchically arranged mountains were supported by the observation of a "cover" of secondary and tertiary rock formations in the Harz massif in Germany (the secondary mountains of these authors) leaning against a Hercynian basement (the primary mountains), which imposes its structural characteristics and controls the present relief.

Whoever refused to accept the explanation of the origin of mountains that conformed to the events described by Moses in Genesis needed at least to justify the hierarchy of their shapes. Buffon used this approach with some success.

## Buffon

Georges-Louis Leclerc, (chevalier and thereafter) comte de Buffon (1707–1788), was born at Montbard (Côte-d'Or, France). His father was councillor at the parliament of Burgundy, and the future naturalist studied law in order to follow in his father's footsteps. However, he had but little inclination for magistracy, and during a trip with a young English friend he discovered nature and decided to become a naturalist.

His final decision to study natural sciences was nevertheless due to an unexpected proposal. In 1739 the director of the Jardin du Roi died suddenly and the position was given to the young Buffon, known only for his translations of some English scientific works.

The royal garden of medicinal herbs had been created a century earlier by Guy de la Brosse, physician of Louis XIII, who did research in human anatomy and medicine as well as botany and chemistry (Rouelle taught there after 1742). The director was appointed by the king and chose professors in turn. At first, the director was the king's physician, but later on scientists were usually preferred.

### A Natural History

Buffon took care of the administration of the Jardin du Roi, which he enlarged considerably (he was a clever businessman). He also began to write a monumental work: his *Histoire naturelle* in forty-four volumes, treating in succession natural history in general and in particular (fifteen volumes, 1749–1767); the history of birds (nine volumes, 1770–1783); followed by seven volumes of supplements to natural history (1774–1789); the natural history of minerals (five volumes, 1783–1788); of quadrupeds (two volumes, 1788–1799); of fish (five volumes, 1798–1803); and finally the history of cetaceans (one volume, 1804, edited by Lacépède).[1]

It may be noticed that this masterpiece is incomplete because Buffon treated only higher vertebrates. However, his interest in those was exclusive. Intolerant toward any contemporary who might overshadow him, he professed nothing but contempt for insects. This allowed him to neglect the work of René-Antoine Ferchault de Réaumur (1683–1757), the famous insectologist who was twenty-four years his elder.[2] If Buffon confined himself to animals closer to *Homo sapiens*, it was perhaps also because he did not want to venture into scientific fields where it was necessary to use a system of classification. Indeed, he might have been afraid of having to acknowledge the superiority of Carl Linnaeus in that field.

The mutual feelings of contempt between the two illustrious naturalists are revealed by the name, *Buffonia*, that Linnaeus bestowed, with his compliments, on a particularly disgraceful plant. Buffon in turn criticized Linnaeus's system of classification, which was based on the number of stamens in the flowers of a species. In the introduction to *Histoire naturelle* he said ironically: "To recognize a tree or a plant, one must carry along a microscope. . . . This large tree that you see is perhaps only a burnet [small rose bush]; one must count the stamens to know what it is; and since its stamens are often so little that they cannot be seen with the naked eye or with a hand lense, one needs a microscope."[3]

The two men had indeed completely different attitudes toward natural history. Linnaeus spent his life classifying things in what was, according to him, the easiest possible way, at the same time trying to approach a natural order. Buffon, on the other hand, showed his contempt for systems when, after his study of humans, he began that of quadrupeds with the horse, "man's most noble acquisition," followed by the dog "because, as a matter of fact, he usually follows the [horse]" whereas the zebra "is little known to us."[4] Classification of animals by their friendliness to humans is of course not following Linnaeus at all, and this could only widen the rift of incomprehension between the two men.

## Theory of the Earth

To return to the history of geology, I must mention that Buffon was interested in the history of the earth at two different points in his career: first, at the beginning of *Histoire naturelle,* where his "first discourse on how to study natural history" is followed by the "second discourse on the history and the theory of the Earth," a theory based on a series of "proofs," which take up the largest part of volume 1 of the work; second, when he published *Suppléments à l'histoire naturelle.* There he added a supplement to his theory of the earth. This new text, entitled *Des Époques de la nature* (Of the epochs of nature), was so well structured and provided such pleasant reading that it was reprinted separately many times and thus became the best-known part of Buffon's whole work.[5]

Buffon's two views on geology were written thirty years apart. The "second discourse," dated 1746, was published in 1749, at the same time as the next two volumes, which contained, most notably, the natural history of human beings. The *Époques* and the additions to the theory of the earth were first published in 1778, in volume 5 of *Suppléments.* Geological knowledge had evolved tremendously in the meantime. For the classification of mountains, for example, works by Lehmann, Rouelle, d'Holbach, and others were now available. Where in 1749 Buffon had made no distinction between classes of mountains, in the additions and in *Époques* three decades later he corrected his earlier statements.

## Plutonism

Buffon made these corrections in a very personal and rather original way. Contrary to the neptunists Lehmann and Rouelle, who believed that the first two classes of mountains were of aqueous origin, Buffon said that the oldest mountains were of igneous origin. According to him, the globe had gone through a molten stage, and during its slow cooling the molten matter became solid and produced irregularities in the earth's crust, which are now high mountains.

If we followed the traditional vocabulary used at the end of the eighteenth and the beginning of the nineteenth century, we might call Buffon's a plutonist theory and contrast it to the prevailing neptunist ones. In fact, the first controversy among geologists in the late eighteenth century concerned the origin of **basalt**—today known as a volcanic lava, that is, an **igneous rock,** but which was at first believed to be a **sedimentary rock.** The two opposing sides in this controversy were originally called neptunists and vulcanists: neptunists if they believed in sedimentation, Neptune's domain; vulcanists if they argued for the deep upwelling of lava, Vulcan's domain. Later on, the term *aqueous origin* was used to describe the formation of rocks in water by precipitation, and *igneous origin* meant solidification of molten magma. Plutonism, however, as accepted by Hutton (see chapter 9), was a different concept. He believed that not only basalt, but also **granite** was of igneous origin.

According to the later terminology, Buffon was a plutonist because he believed in the igneous origin of granite, the main constituent of primitive mountains. Moro, on the contrary, would have been a vulcanist. Nevertheless, Buffon's plutonism was not the same as Hutton's because Hutton stated that granitic magma is formed periodically if not continuously, whereas Buffon restricted its formation entirely to the past. Using the prefix *paleo-* (old), we might describe Buffon's ideas as paleoplutonistic and draw an important difference between the two authors: Hutton was uniformitarian whereas Buffon was not.

## From Cyclicity . . .

Nevertheless, Buffon's "Théorie de la terre" called for causes as close as possible to those which exist in the present; namely, for "effects which occur daily, movements which follow each other and are renewed without interruption, processes which are constant and always repeated."[6] In 1746 Buffon still believed in the cyclic return of the same events: if the "waters of the ocean" once covered our lands, the "waters of the sky," which presently erode mountains, "shall return this land to the sea which will absorb it over time, leaving other continents exposed."[7]

At the same time he was also a true neptunist because he attributed all mountains to submarine deposits. When he became a plutonist in 1778, geological phenomena were evidently no longer repeated. Did not the title of the new work emphasize the irreversible sequence of events? Every "epoch" was marked by a particular phenomenon, which made it unique and hence different from other epochs.

This point, which corroborates a thesis proposed by a modern historian of religion, Mircea Eliade, warrants some remark. Based on studies of primitive societies, Eliade stated that periodic ceremonies for the expulsion of devils, diseases, and sins often coincide with a new year because these rites are supposed to abolish time, to return to the act of creation of the world. By returning to the origin, such societies and individuals in them attempt a kind of regeneration in order to escape historic time. In this philosophical framework, human existence and personal events are regarded as a Fall that must be cleansed by a perpetual return to the origin. However, this allows no place for history. History can only be defined as the invasion of individual and irreversible actions into this system of mythical creation and re-creation.[8]

Although this thesis does not apply directly to the history of the earth, from which the individual is absent, the confrontation between irreversible processes and the cyclic return of the same phenomena has always played an important role in that history. Indeed, the cycles of the Stoics represented a means to escape history, and the Aristotelian cycle (more muddled because it was more directed toward a permanent equilibrium than a periodic return) did not leave more room for a sequential history than the one proposed by Buffon in his Époques.

Individuality, a characteristic of human history, has its equivalent in the role of randomness. As we pointed out earlier, history is based on some kind of dialectic between randomness and laws. In human history, are not unpredictable individual actions leading to randomness? By analogy, we could perhaps say that the past of the earth becomes truly historical when it is so complex that it cannot be predicted entirely or deduced completely from natural laws.

## . . . To Irreversibility

According to such a historical definition, Buffon has indeed introduced an important element into the earth's history by making terrestrial phenomena irreversible. But in turn, he tended to erase randomness for the sake of postulating rigorous laws that allowed him to neglect the archives of the earth.

The leading factor in the sequence of epochs in his "Époques de la nature" is the cooling of the globe.[9] Buffon proposed that there have been seven periods, just as many as there are days in the week—that is, according to Genesis—which correspond to stages of cooling.

During the first epoch, the earth and the planets were formed. Repeating a theory stated earlier in his "Théorie de la terre," he assumed that a comet had brushed against the sun and detached a gush of molten matter, which divided into as many masses as there were planets in the solar system, namely six (Uranus was discovered shortly after publication of "Époques de la nature").

During the second epoch, matter became solid and formed the interior rocks of the earth as well as the large vitreous masses at its surface. These masses are in fact mountains of igneous and metamorphic rocks; Buffon believed that crystallization of granite was caused by **vitrification.** He thought that irregularities at the surface of the earth were similar to those formed toward the surface of a molten mass of metal or glass undergoing cooling: "At the surface are formed holes, waves, uneven parts; below the surface are formed empty spaces, cavities, swellings which may give us an idea of the first irregularities which developed at the surface of the Earth and of the cavities inside." [10]

During the third epoch, water—which until then had existed as vapor in the atmosphere—condensed, and continents were covered with water. The proof of the presence of the sea is provided by shells and other marine organisms, which are found up to 1,500 *toises* (1 toise = 1,949 meters or 6,395 feet) in the Alps, and even up to 2,000 *toises* in America [South America]. This sea covered the vitreous primitive mountains, first with sedimentary layers of clay, and then of limestone. Only the summits of high mountains, where shells are not found, towered above this universal ocean.

It is during the fourth epoch that the waters retreated and volcanoes became active. Dry elevated land gave shelter to plants, which after their death produced coal and pyrite. The effervescence of pyritous and combustible materials ignited fires, that is, volcanoes. Volcanoes occur close to seashores, he explained, because effervescence requires water and the seashore allows the necessary collision of a large volume of fire against a large volume of water. During the lowering of the ocean, volcanoes spread toward new seashores and former craters became extinct.

Buffon explained the extension of land by the former presence of subterranean caverns, which were some sort of giant bubbles formed during the cooling of the globe—at the same time and in the same manner as the irregularities at the surface of the earth. When the ocean undermined the arches and pillars of these subterranean caverns, their roofs caved in and the ocean invaded these "valleys," which led to a lowering of sea level.

During the fifth epoch, elephants and other animals, living today in warm climates, inhabited the northern countries. The cooling of the earth being gradual, polar areas were the first to reach temperatures compatible with life because it rained there earlier than at the equator. For a long time, polar areas had the same temperature as today's equatorial areas, whereas the original equatorial regions were still deserts with temperatures near the boiling point. Hence, polar areas were the first to acquire the fauna and flora that live today only in equatorial countries.

## Lost Species

Buffon wrote that during the fifth epoch the old and the new continents were adjacent because elephant bones had been found in Russia, Siberia, and Canada, "close to the Ohio river." Furthermore, the large molars of the hippopotamus and of a huge animal—whose species has been lost—had come from Canada, whereas other, virtually identical, molars had come from Siberia.[11]

Of course, what Buffon called an elephant was in reality one of its ancestors, the mammoth; and the "huge animal whose species has been lost" was in fact a mastodon, a species also related to the present elephant. What Buffon believed to be the molars of a hippopotamus belonged to this species too. Moreover, the mastodon was not as large as modern elephants and thus not as "huge" as he believed. Furthermore, he did not understand the ecology of these animals. The fact that elephants now live only in equatorial countries does not at all imply that all its ancestors lived in similar climates. It is known today that the mammoth, protected by its thick, hairy skin, withstood cold climates and spread to warmer regions only during periods of cooling corresponding to **Pleistocene** glaciations.

However, he was correct when he talked about "lost species," adopting the term used two centuries before by Palissy (chapter 5). At the end of the eighteenth century, lost species, which had intrigued Hooke and Leibniz after Palissy, became a constant preoccupation for naturalists. Paleontology was in the process of being born.

Buffon stated that during the sixth epoch, continents were separated. Thereafter came an epoch during which the power of humans was added to that of nature.

The last epoch was actually added during the typesetting of the book; according to Jacques Roger, the original manuscript of "Époques de la nature" contained only the first six epochs.[12] This was closer to the six days of Creation of the world, which had earlier inspired Steno to divide the history of Tuscany into six stages.

## Conflict with the Church

Buffon did not follow the story of Genesis as faithfully as Steno. When he referred to the Bible or to God, he was merely being cautious. In the manuscript, he did not hesitate to say, "Analogy, monuments, and even tradition show that the human species originated at the same time as all the other species," which rendered the distinction of a seventh epoch useless. However, when he repeated this bold statement in the final text, he ascribed it to devil's advocates and preferred to conclude: "Is it possible to doubt that we differ prodigiously from animals because of the divine ray which the Sovereign Being sent our way?"[13]

We may doubt the author's sincerity in such statements, knowing that he considered such language pure "mockery." At least this is what was reported by Hérault de Séchelles, who visited Montbard in 1785. Buffon told his visitor about the difficulties he had had with the Sorbonne with respect to the publication of the first three volumes of *Histoire naturelle* in 1749 and admitted cynically, "I did not make any difficulties in providing all the satisfaction that it [the Sorbonne] might have wished: it is only a mockery, however, men are stupid enough to be content with it."[14]

It is true that his behavior with the Sorbonne was a model of feigned humility. He showed his good will by publishing the letter of his censors and his answer at the beginning of volume 4 of his works: On January 15, 1751, "the deputies and the syndic of the faculty of theology" pointed out fourteen propositions made in the published volumes of *Histoire naturelle* that were "contrary to the beliefs of the Church"; the first four concerned his theory of the earth. Without going into any arguments, he calmly answered that he had "no intention to contradict the text of the Scriptures," and that consequently he "abandoned all which in [his] book concerns the formation of the Earth, and in general all that might be contrary to the story by Moses."[15] However, thirty years later, in the "Époques de la nature," he developed and enlarged his theory that the solar system had been

formed by a collision with a comet. He thus ignored his earlier pre-
cautions and stressed instead that the six days of Creation "are
merely six spans of time" whose "duration cannot be determined by
the historian of religion."[16]

## The Age of the Earth

Liberties taken with the text of Genesis and its literal reading made it
possible for Buffon to extend the age of the earth to some 75,000
years. He estimated that it took 2,936 years for the consolidation of
the earth, that the third epoch began "thirty thousand or thirty-five
thousand years after the formation of the planets," and that the fourth
finished 45,000 to 60,000 years after that date.

Compared to the 6,000 or 8,000 years calculated according to the
Bible, these time spans may appear immense and audacious. However, even in these numbers Buffon hid his true belief: his manuscript laid down 3 million years for the entire time span, including
117,000 years for consolidation alone.[17] Why was he afraid to publish
these figures? Without any doubt, the Church was just as shocked by
hundreds of centuries as it would have been by periods fifty times
longer. It is possible that Buffon wanted to convince the public at
large; to avoid stirring up general disbelief, he reduced the length of
geological time. In his manuscript he wrote, "The more we extend
time, the closer we shall be to reality." However, he immediately
added: "Nevertheless, we must shorten it as much as possible to con-
form to the limited power of our intelligence."[18] Was Buffon address-
ing a large public whose "limits of intelligence" he tried to spare?

Moreover, he did not care to be mistaken for one of those liberal
authors, such as Maillet or the "Turkish spy," who had proposed extremely long time spans. He was, after all, the director of the Jardin
du Roi, and his Histoire naturelle was published by the Royal
Printer. There were certain things that his position would not allow
him to write.

This is certainly the reason why he tried to save face when the
Sorbonne squabbled with him. This is also why he remained silent
when attacked by his contemporaries. Indeed, a higher civil servant
can hardly engage in polemics. Moreover, his extreme vanity made
him believe that he ranked among the greatest geniuses that ever
lived on earth, in the company of Newton, Bacon, Leibniz, and Mon-
tesquieu. This high opinion of himself prevented him from answer-

ing mediocre adversaries. Furthermore, he was saving time: his huge work should not suffer any delay because of vain quarrels.

## Borrowed Ideas

It is possible that Buffon could not acknowledge all his sources. For instance, his theory of the earth published in 1749 included ideas on the formation of mountains by ocean currents and on shells found in high mountains—ideas mentioned earlier by Maillet, whose *Telliamed* he had certainly read. In "Époques de la nature," Buffon borrowed ideas from Leibniz and contemporaries such as Lehmann, authors whom he could have cited although he neglected to do so. He had also borrowed from Nicolas-Antoine Boulanger, whose manuscript entitled "Anecdotes de la nature" he owned for some time.[19]

Boulanger was an author with a smack of heresy. Born in 1722, he died prematurely at the age of thirty-seven. He is known for two posthumous works, *L'Antiquité dévoilée* and *Traité du Despotisme Oriental*, both violent critiques of religion.[20] He was also the author of articles written for the *Encyclopédie*, in particular the one on the Flood. Buffon had always refused to become involved with Diderot's and d'Alembert's enterprise.

In reality, Boulanger's thesis is very different from Buffon's because it proposed a cyclic evolution of the earth in which the alleged creation of the world was only the last revolution to which the earth was subjected. Nevertheless, Boulanger had studied in great detail the meanders of the Marne River, and his description is very close to the one given by Buffon at the end of the fourth epoch, where he wanted to prove that valleys were carved by the retreat of marine waters rather than by the action of rivers flowing there today.

In fact, Buffon borrowed only the descriptive aspects, which were very tempting since Buffon was familiar with the same area as Boulanger. All this does not diminish the originality of "Époques de la nature," nor Buffon's attempt at an irreversible history of the earth. However, were all the features of a history of the earth really combined in the "Époques"?

## An Aborted History

It is not geologic time that is missing in Buffon's theory (though he was forced to shorten the time span). Nor is it irreversibility, because the history is already complete. Buffon's theory is flawed because he got rid of the true archives. His successive epochs represent mostly reconstructions based on his ideas alone. Buffon's own observations are limited in number, and their impact is lessened by his attempt to explain these observations by some law of nature.

For instance, in the third epoch, he said, "the formation of clay seems to have preceded that of shells because the first action of water was to transform scoria and glass powder into clay." He has just laid down as law that clays preceded limestones. "Thus," added Buffon, "layers of clay were formed some time before limestone layers, and one can observe that these deposits of clayey material preceded those of limestone because almost everywhere limestone rocks rest on clay." [21] In other words, Buffon's observation of limestone resting on clay comes only at the end of his demonstration; he gives it as factual proof of a law that was first established by reasoning alone. His order of deposition of layers is therefore not based on empirical knowledge; it is deduced first of all from natural laws. Without returning to the Cartesian method, Buffon searched for a law that would explain observations and would allow him, after its establishment, to do without them.

Archives thus became irrelevant. The historical method, according to the definition given earlier, was discarded. The order of events became accessible to the effort of reasoning, and reconstruction of the past remained purely conjectural. Only unpredictable observations needed observation and research of documents. Buffon did indeed accept the possibility that here and there clay was deposited locally upon limestone, contrary to what he had forecast. Observations outside of Burgundy would have shown multiple examples of clay above limestone because Jurassic rock formations of the Paris Basin consist of alternating beds of limestone and clay.

He was not, however, particularly interested in meticulous studies of series, and even such observations would certainly not have influenced his convictions. Why stop at what seems to be random when one has a general law?

In reality, there was no such thing as a history of the earth just yet. Nevertheless, it was ready to emerge. Buffon's concept of irreversibility gave it a better chance to start than the cyclic systems of the ancients. However, the search for a law of evolution remained an obstacle to the effective use of the archives of nature. Furthermore, in order to reach these archives, it was necessary to observe carefully the superposition of the earth's layers. This was done only in the last ten years of the eighteenth century, when industrial needs required such knowledge to exploit mineral resources below the earth's surface.

# Chapter 8

# At the Service
# of Industry

## The Industrial Revolution

Buffon's interest in mineralogy was at least partly practical, because
he was the owner of foundries at the hamlet, close to Montbard, from
which his name originated. In the eighteenth century, the study of
the earth's layers was thus no longer the sole domain of theologians
talking at length about the Deluge and the Creation of the world. Indeed, it became of interest to industrialists who wanted to improve
mining in their country.

This interest was not entirely new. At the beginning of the sixteenth century, Agricola (1494–1555) was interested in both mineralogy and metallurgy, and his 1546 work, *De re metallica* (Of
metallurgy), described the state of the art in geology, mineralogy, and
metallurgy of his time. However, in the eighteenth century, mining
became much more important because of increasing needs related to
the industrial revolution.

England was the leading coal producer, with huge production increases. In 1700 the annual yield was 2.5 million tons of coal in the
entire country. In 1770 it was 6 million; in 1800, 10 million; and in
1830, 30 million tons per year. Coal production thus increased 12-
fold in 130 years.

France and Germany lagged far behind. It was only after the Treaty
of Utrecht (1713) that the French mining industry developed. In 1744

it was decided that mines could be granted to persons other than the owners of the land, a decision that greatly enhanced the rise of the industry. In 1747 the École des ponts et chaussées (School of Civil Engineering) was founded and began to train some of its students each year as mining engineers.

## Mines

H.-L. Bertin, in charge of the French technical ministry for the supervision of both agriculture and mining, proposed in 1765 the creation of a school of mines. It opened in 1783 but lasted only five years. It was reopened in 1795, during the French Revolution.

At the time of Bertin's suggestion of a French School of Mines (in 1765), Germany founded its Bergakademie in Freiberg, Saxony, in response to the same general demand for teaching the art and science of mining. The location was not a random choice; for a long time the miners and workers of Saxony had enjoyed the reputation of competence in the field of metallurgy. Ore deposits had been extracted there for a thousand years. Agricola, the "father of mineralogy," was from Saxony.

To exploit ores it was, of course, necessary to know how rocks visible in **outcrops** extended down below the surface. Miners had some empirical knowledge handed down during centuries of practice. However, when industrial needs increased dramatically, it became urgent to rationalize research methods. Beginning in the middle of the eighteenth century, geologists made great strides in their knowledge of the order of rock layers. Exploration of mountains (chapter 6) had already led to the distinction of primary and secondary rock formations. Most of the naturalists who advanced this knowledge were working in the field and wanted to improve the use of mineral resources. They did not hesitate to take advantage of the miners' empirical knowledge.

For instance, Lehmann, whose recognition of three classes of mountains we have already mentioned (chapter 6), first studied medicine. When practicing medicine at Dresden, in Saxony, he soon became interested in mining and metallurgy and studied the arrangement of layers in that area, in particular, in the Erzgebirge. He also visited the Harz, another mining region to the north, where mining had begun in the year 745. In 1754 Lehmann was appointed mining councillor or Bergrat, and he served as director of a copper mine.

Arduino, who had recognized two orders of mountains at the same time as Lehmann, was also actively involved in mining. He even worked in a mine in his youth before becoming an expert on mining (as well as on problems in agriculture and industry).

## Subdividing Secondary Mountains

A simple distinction between two (or three) classes of mountains was not good enough for miners. It was within each of these classes that the order of superposition of layers had to be established. It was most important to locate the oldest class since it was known to be richest in metals; however, the horizontal secondary layers were easier to classify and were well understood before all others.

Lehmann was the first to realize the importance of the order of strata and to make use of the accumulated knowledge of quarrymen and miners. However, these people distinguished strata with such fine points, often too precise for geologists, that certain subdivisions were of purely local interest. Lehmann knew how to adapt this empirical knowledge to the demands of the emerging new science. He divided the stratified rocks (*Flötz-Gebürge*) that surround the Harz and which always lean against primitive mountains (*Gang-Gebürge*) into red beds consisting of coarse sand and pebbles, followed by beds of coal, shales, limestones, and finally by "salted springs."[1] In modern terms, this sequence represents the **Permian,** a period ending the Paleozoic, the sequence consisting of red conglomerates, sandstones, shales, and coal beds, followed by black shales, limestones, and finally **evaporites.**

Georg Christian Füchsel (1722–1773), physician to a prince of Thuringia (a former state in central Germany), improved upon Lehmann's classification. In 1761 he published in Latin a history of the land and the sea based on the investigation of his province (*Historia terrae et maris*).[2] Above the layers of salt and gypsum, he recognized variegated sandstones and fossiliferous limestones (called *Muschelkalk*, or shelly limestone), which form two of the rock formations in the first period of the second era (Mesozoic), today known as **Triassic.**

Füchsel's superiority over Lehmann lies in his understanding that layers were formed successively over long periods of time, and not during the unique event of the Deluge. He said, "In the formation of deposits, nature must have followed present-day laws: every deposit forms one stratum, and a series of strata of the same composition represents a formation or an epoch in the history of the globe."[3]

Füchsel's division of epochs in the earth's past, based on the investigation of "formations" of strata, opened the door to the naturalist whose name became most closely linked with the classification of strata (perhaps at the expense of his predecessors): Abraham Gottlob Werner (1749–1817). Like Lehmann and Füchsel, Werner was a typical product of the young academy at Freiberg, where he studied from 1769 to 1771 before becoming a professor in 1775. His ancestors had occupied important positions in the industry of mining and metallurgy of that area since the beginning of the sixteenth century.

## Werner

Werner began his scientific career with a work on mineralogy, *Von den äusserlichen Kennzeichen der Fossilien* (On the external characters of minerals), which was received with enthusiasm all over Europe and was translated into many languages.[4] Mineralogy was then developing as quickly as historical geology and was supported by chemistry, another fast-emerging science. However, Werner's goal was purely practical: he based his classification on the external characters of minerals such as color, cohesion, external shape, luster, fracture, transparency, hardness, specific weight, smell, and so on. He put little emphasis on crystal shape and chemical composition of minerals, although his contemporaries in France, first Romé de l'Isle and then René Just Haüy, were showing that the external shapes of crystals resulted from a small number of crystalline systems (as they were later called, although Haüy named them the "integrant molecules").[5]

## Geognosy

Throughout his life, Werner continued to elaborate on his work of classification. A new science was born, which Werner called *geognosy* after a term used earlier by Füchsel.

Jean-François d'Aubuisson de Voisins, Werner's student, said that "in geognosy, rock formations and superposition are considerations of first order. Consequently, mineral masses have to be arranged and classified, as much as possible, according to their order of superposition or their relative age."[6] This shows that the Wernerian method addressed two concerns at once: one taxonomic and pedagogic, the other scientific. By classifying and naming rock formations according to their order of superposition, miners could learn about the structure of the subsurface. At the same time, the system made the scientific point that this order accurately represented the relative age of layers.

The second point is, of course, a simple application of the principle of superposition stated by Steno 120 years earlier—a well-established fact in science. However, when the rule was applied in practice, it had to take account of some exceptions because it originally assumed that layers had not been disturbed after deposition. If a series has been overturned during folding (**recumbent fold**), the lower layers are the youngest. Similarly, if a rock mass has been thrust over another one (**overthrust**) during compression (chapter 15), the superposition of layers no longer follows the rule, at least not at the site of the "abnormal contact."

Moreover, the principle of superposition does not apply to igneous rocks when magma intrusions cut through older rocks. Based on the principle of cross-cutting relationships, any sedimentary rock that is cut by an igneous intrusive body is older than that intrusion.

Werner refused to accept either tectonics or **magmatism**; he believed that layers preserved the position they had when they were deposited. The present superposition was therefore one of deposition. Furthermore, according to him granite was not an intrusive rock but a primitive deposit, so plutonism did not exist.

According to Werner's view in *Kurze Klassifikation*, the order of deposition of the principal rocks made it possible to distinguish four classes of rocks (or mountains): primitive (*Uranfängliche*), in layers (*Flötze*), volcanic, and alluvial. Around 1796 he inserted a new group between the first two classes: the transitional mountains (*Übergangsgebirge*), which included tilted and fossiliferous late Paleozoic layers resting on primitive igneous and metamorphic rocks. Each of these classes was naturally divided into various categories.[7]

Stratified mountains, for instance, were subdivided into twelve formations, beginning with the period today known as **Devonian**. At the top were the salt-bearing layers of the Zechstein (an evaporite-bearing formation), and thereafter those of variegated sandstone and fossiliferous limestone of the Triassic, and so on. The series ended with basalt layers and other volcanic rocks Werner believed to be sedimentary, followed by chalk of the Cretaceous, which he placed, strangely enough, on top of the series, above the gypsum of the Paris region, which is actually younger.

## Long-range Correlations

It was not sufficient to reject igneous phenomena and tectonic movements to establish a "geognostical column" of "formations," or rock layers. A problem, today called **long-range correlation,** arose instantly. When drilling in the ground at some point on the earth, we find superposed layers, which we assume to be increasingly older according to the principle of superposition. Of course, eighteenth-century observers did not have any drilling equipment at their disposal. However, if layers are tilted in a regular fashion, it is possible in the field to trace successive strata in their order of increasing age without drilling. That is, the naturalist would seek a superposition in the field that was similar to a stack of cards that had been tilted and slid apart as the cards were spread on a table. Ever since the middle of the eighteenth century, geologists had known that the edges of mountains offer this arrangement and thus show the oldest rock formations.

However, if the same exploration were made in another country, or on a different continent, who could guarantee that the same layers would be found? We touched on this problem in chapter 6 and noted that it was partially solved in favor of a universal order of superposition of layers. As a contemporary of Saussure and Ramond, Werner, too, wanted to find the same arrangement in various areas. More precisely, once the study of several mountains had shown a similar arrangement of large masses of rocks, Werner accepted the hypothesis that this similarity also existed at the level of layers. In other words, he believed that a particular type of rock was characteristic of a particular time. It was merely necessary to observe one area carefully in order to know the universal order of all layers of the earth. Werner, of course, chose to study systematically his native country.

## The Need for Classification

Werner was by nature always tempted to classify, to put things in order. Georges Cuvier compared him to Linnaeus. It is true that the two naturalists had the same concern for nested classifications. Like the Swedish botanist, Werner distinguished classes (of mountains), which he divided further into distinct kinds and varieties. In his definition of divisions, Werner, like Linnaeus, was searching for "characteristics which were constant" rather than those that varied in a disorderly fashion.

However, in both cases, the enterprise was superhuman because the authors wanted a simultaneously practical and natural classification. Werner, for whom natural order was the order of deposition, ended up placing *all* formations of the world in the *same* column. Two difficulties appeared immediately. One was a question of principle, the other of facts.

Werner believed that *general* and *partial* formations existed and that the former were "produced by a general cause" and the latter by "particular and local causes." True, partial formations could be dated in relation to general formations. However, when local formations occurred far apart (i.e., when they were not in contact with one another), they could not be dated by comparing one to the other because the principle of superposition could not be applied. Their relative age could not be determined even if two distant partial formations occurred between two general formations. However, since the Wernerian system was based on superposition and since the system could not visualize a lateral transition between formations, partial formations of the same age were simply superposed in one way or another. This was the first difficulty.

The second difficulty was worse. Having at one's disposal only a limited number of local series, formations are, of course, more or less in the same order. Since he believed that numerous universal formations covered the entire earth like layers of an onion, Werner's ambition was to offer a single column that included rock formations from the entire world. Alexander von Humboldt (1769–1859), one of Werner's most famous students, traveled all over South America trying to establish such an order of layers.

The system now showed its weaknesses because a given formation could have only one place in the column. As Aubuisson observed, it seemed as if "what had already been put in order, was plunged into confusion again." Granite, which was believed to be older than gneiss, reappeared above gneiss—even above **micaschists,** if not on top of the even younger **phyllites.**[8]

Today, the reason for this confusion is obvious. The distinction of the two classes of mountains in Germany corresponds to reality: the superposition of a younger sedimentary cover over the older

Hercynian basement. But divisions inside these two units are less obvious, especially in the Hercynian basement, where intrusive granites cut across gneiss and micaschists so that the same "formations" appear at different levels of the "column." Things are less complicated in the sedimentary cover because there is neither intrusion nor metamorphism. But there, superposed layers are dated by their fossil fauna and flora and not by the nature of rocks, that is, their lithofacies.

This is where the Wernerian system failed. Today, it is common knowledge that epochs cannot be distinguished by a single and unique deposit. Indeed, everybody can observe that layers of different kinds are being deposited at the surface of the earth. Here it is sand, there it is argillaceous mud, elsewhere calcareous ooze, and so forth. Why not accept what is, so to speak, written on the wall?

## Anti-Uniformitarianism

The school of Freiberg as well as all neptunists refused to accept this evidence. They were not at a loss for arguments when they asked: Why should we believe that in the past nature was identical to the present? Convinced that granite was formed as a deposit, they argued that nothing is happening today that produces granite. They reasoned that the primitive earth had been surrounded by a universal ocean, which held in solution all the materials that were thereafter precipitated successively.

They believed that what can be observed today does not offer any clues about these ancient formations. J.-A. Deluc said, "The residue of the *primordial liquid* which is the sea no longer forms *layers of minerals.*" Ancient layers were formed, he said, by "primordial causes which exist no more." Their formation can be explained only when "going back farther than present-day causes."[9]

We find here a serious objection to uniformitarianism. It is, however, a paradox that Deluc, who first used the phrase present-day causes, or more precisely, "causes now operating," actually wished to refute uniformitarianism. In other words, he first used a term characterizing the method governing modern geology in order to refute that method. I shall return to this point. For the time being, I merely want to stress that the study of the archives was, once more, at a standstill.

The neptunists posited that the order of precipitation of materials in the universal ocean "born from chaos" was not random. They stated that the first deposits consisted of large transparent crystals. With increasing movement of the waters, crystals became smaller and less distinct, drowned in an opaque loose material. Furthermore, the waters of the ocean lowered progressively (by evaporation, according to Werner) and the highest primitive sediments, deposited on top of elevations of the primitive crust, began to emerge. They provided "mechanical" deposits, which mingled with the primordial "chemical" precipitations, contaminating them with completely opaque materials: crystalline granites were replaced by sandstones and conglomerates.

Neptunism was, of course, not as homogenous a doctrine as given in this all too brief summary. Indeed, on the one hand, Werner was not anti-uniformitarian, and on the other, Deluc, following Füchsel, introduced in his system collapses that interrupted the regular course of events.

Nevertheless, despite important differences among themselves, neptunists adopted many common themes embodied in the main doctrine. For instance, many accepted the position that the history of the earth is regressive. Ever since Celsius and Linnaeus it was believed that the earth was losing its water. The Wernerian system introduced the idea that the ocean also became impoverished in dissolved materials. In fact, for two reasons the earth had as yet no real history.

## A World without a Future

On the one hand, the neptunists believed that, if there were a law of evolution, it would be sufficient to study one region in order to know the arrangement of layers throughout the world. Thus the problem posed by the two classes of mountains returned in a new form—and required no long-term investigation. To naturalists it seemed that the search for archives had reached its limit. Science seemed on the verge of its own completion; the mere statement of a general law seemed to be a dispensation that excused travelers from further explorations.

On the other hand, the neptunists' arguments implied not only that science was coming to a standstill, but also that the history of the earth ended equally abruptly with the present world. The planet thus had a past but no future. It was a common belief that the forces that had formed the earth were being depleted. One finds here again the idea cherished by Buffon, which actually goes back to Lucretius;

namely, that nature loses its power of reproduction and that its productions are becoming more and more puny. Working under such
assumptions, Deluc or Werner could imagine a formation but not a
history of the earth. Such a history had to overcome these two obstacles and state the following propositions:

> Each region has its own unique history, and the history of the
> world consists of the juxtaposition of local events. Our research
> in the archives is never finished. Geology means fieldwork. No
> general law of evolution will ever save us from traveling around
> the world, hammer in hand.

> The present moment is only one point on the uninterrupted thread
> that unravels the history of the earth, starting with the planet's
> first revolutions. The forces that fashioned it are still in action,
> and they shall modify the earth, or rather they are changing it
> imperceptibly in front of our eyes. However, to take into ac-
> count such slow yet always active causes, we must accept enor-
> mous durations of time.

In order to solve the second problem, that of the vast time spans,
which no one, not even Buffon, had proposed, geologists needed
a mechanism capable of ensuring earth processes at a uniform rate.
Hutton was to propose this mechanism. The first problem could be
solved only when geologists had learned how to use this mechanism
for each region—or rather had learned to recognize locally the effects
of internal processes of the earth. Only then would it make sense to
return to the study of the geological archives.

## Desmarest on Volcanism

Nonetheless, quite some time before arriving at the stage of understanding the major internal processes of the earth, some naturalists
had tried to undertake the regional approach. In other words, they
had used the archives provided by their microcosm to sketch a pretty
valid history.

In 1779, shortly after the publication of Buffon's "Époques de la
nature," Nicolas Desmarest (1725–1815) wrote an article called "On
the Definition of Some Epochs of Nature by the Products of Volcanoes and on the Use of These Epochs in the Study of Volcanoes." [10]
The modest title shows the limitations of the work: only "some"
epochs were considered. Indeed, the author did not go back to the
origin of the earth but only to relatively recent volcanic events. More-

over, his work concerned only volcanism. However, there are times when the most scientific approach to a problem is to rein in ambition, to narrow the field of investigation to a more limited but more detailed study.

Compared to Buffon's, Desmarest's method of investigation was new. Instead of presenting the history from beginning to end, he reversed it, starting at the present and going backward by induction based on documents. Hence, this memoir was far from plagiarizing Buffon's book; it was, in fact, just the opposite. Desmarest first read his essay in 1775 at the Academy of Sciences in Paris, before Buffon's "Époques de la nature" was published.

The interest in volcanism gave Desmarest another priority because he was the first to understand that basalt is ancient lava. Since 1752, Guettard, friend and student of Réaumur, had observed that the mountains in Auvergne were nothing but extinct volcanoes. However, he separated basalt, found abundantly in that region, from volcanic lava.[11] In 1763 Desmarest confirmed Guettard's observations. However, he went further and declared that basalt was also volcanic.[12] His discovery started the debate between vulcanists, who followed him, and neptunists, who continued to believe that basalt was a sedimentary rock.

The most famous neptunist, Werner, remained adamant until his death, despite the increasing number of proofs to the contrary. His disciple Aubuisson de Voisins remained respectful of the prejudices of the great man and waited until Werner's death for the publication of his observations on Auvergne, which finally destroyed neptunistic ideas.[13] One must admire both his loyalty as a student and his freedom of thought, which allowed him to arrive at the truth in spite of the prejudices of his master. Indeed, one measure of Werner's significance is the number of his students who came to dissent from his conclusions but continued to acknowledge the quality of his method of investigation.

## Soulavie

Desmarest did not, in fact, present a complete regional study. Such a study can be found, however, in the seven volumes of the *Natural History of Southern France* by J. L. Giraud Soulavie (1752–1813), published between 1780 and 1784.[14] For the area between Auvergne and Montpellier, he sketched a more complete history than Desmarest had done; he not only recognized the volcanic nature of the

## Coal Mining

The word *coal* first meant charcoal. The use of fossil fuels in Europe, in particular coal, occurred relatively late. Apparently, blacksmiths used coal in coal basins as early as the ninth century. The first coal mine existed at Zwickau, in Saxony. In the area of Liège, Belgium, coal was extracted in the eleventh century. According to some, the French word for coal (*houille*) derived from a blacksmith of that area called Houllos or Hullioz.

In the twelfth century, coal came to the rescue of the forests, which had been overused, and the importance of coal increased at the beginning of the thirteenth century. Thereafter, coal mines were discovered at Saint-Étienne and Newcastle. According to a document of 1315, a citizen of Pontoise loaded wheat to sell at Newcastle and brought home coal.

In the fourteenth century, people began to complain about pollution produced by the burning of coal, and Henry IV of England forbade its use in London. In the middle of the sixteenth century, the king of France, Henry II, condemned blacksmiths who used coal. Nevertheless, coal mining became more and more important. The first mine in the northern basin of France was opened at Anzin in 1734. French coal production reached 250,000 tons in 1789. In 1815, it was 1.8 million tons. In the beginning of the nineteenth century, new needs had to be met. In 1801, Philippe Lebon invented gas lighting. In 1807, Fulton constructed the first steamboat. And the first locomotive ran between Paris and Versailles thirty years later.

Plateau of Coirons, but also included a relatively detailed and precise stratigraphic study. In his stratigraphy he used two novel methods, which later became common practice: the use of a map on which he recorded the various outcrops, and the use of fossils to separate the various levels of "the limestone realm."

In chapter 10, we shall see that Soulavie anticipated works done at the end of the century and the first years of the next. He was nevertheless preceded in that area by P.-A. Boissier de la Croix de Sauvages (1710–1795), who as early as 1749 divided the mountains near Alès, covering a surface of about twenty square kilometers, into ten "chains," which he tried to classify according to their rocks and fossil fauna.[15]

However, before returning to the problem of the beginning of biostratigraphy, let us make a small detour by way of James Hutton's tectonics and plutonism.

# Subterranean Fires

## James Hutton

Histories of geology generally present Werner first and then his Scottish adversary Hutton (1726–1797), although Hutton was born twenty-four years before Werner. The reason lies in the 1795 publication date of the final two-volume edition of Hutton's *Theory of the Earth*,[1] and in the ensuing controversy, which was essentially upheld by his disciple John Playfair (1748–1819).[2] Therefore, the Wernerian system was published a few years before Hutton's.

Hutton was a man of the Enlightenment. He had studied medicine in the first half of the century and defended his thesis the year Werner was born. This thesis partly explains his theory of the earth because it describes blood circulation in the microcosm, that is, in humans (in Latin, *De sanguine et circulatione microcosmi*).[3] Hutton's geological theory was based, in turn, on circulation of matter. He became interested in geology only after he had abandoned medicine and devoted himself to agriculture. Soil is a product of weathering rocks, and soil is necessary for plants to grow. Therefore, mountains have to be eroded to form arable land. However, soil is transported by running water toward the sea; this movement from top to bottom would level continents if it were not compensated for by a movement in the opposite direction that repairs the effects of erosion. The author was so close to the biological model of circulation that he compared the earth to an "organized body."[4]

So the old metaphor of the microcosm was back. Looking at these rather naive proposals of a physician-farmer who deduced uplifting of mountains from the necessity of maintaining land for the perpetuation of the human race, we might ask if no other proposition could have been used to refute geognosy and its observations, which were so useful to miners.

Let us not judge too quickly, for Hutton was also interested in technical progress. What perhaps anticipated his model of geological processes of the earth was his excellent understanding of the steam engine. He was a friend of James Watt, inventor of this engine, and Hutton's geological cycles certainly resulted as much from his interest in the steam engine as from blood circulation in the human microcosm.

## Subterranean Heat

Hutton's system was based on the action of subterranean fire or heat, to which he attributed three effects: induration of sediments, uplifting of strata and formation of mountains, and granitic intrusion in liquid form into layers.

Earlier we called the theory of the igneous origin of granite plutonism and contrasted it to the neptunist theory of the supposed aqueous origin of igneous and metamorphic rocks. It was, however, not the problem of the origin of granite that started the controversy between plutonists and neptunists. Frank Dawson Adams put it this way: "The question as to how it came about that the incoherent sediments laid down in the sea became compacted into solid rocks, was one which presented itself to every observer, but to which the Neptunists and Plutonists gave entirely different answers."[5]

Everyone wondered about the process, today known as **diagenesis,** that changed unconsolidated deposits at the bottom of the ocean or a lake into solid rocks. The simplest answer was to assume a cementation by dissolved substances, which filled the interstitial spaces between superposed particles. This was the belief of neptunists, and it is indeed the process of transforming sand into sandstone. Hutton said, however, that induration occurs sometimes with substances already known to be insoluble (silica in sandstone, for example). He reasoned that a substance therefore becomes liquid by a process different from dissolution, namely, melting by heat. He maintained that fire melts a portion of the sediment by fusing together particles that have remained solid; upon cooling, this portion becomes solid.

From this perspective, the Huttonian thesis naturally seems inadequate. His adversaries were closer to understanding the actual processes of diagenesis. However, at that time, no one could decide between the two answers; to opt for just one explanation seemed unnecessarily extreme. This is why the qualifying (and pejorative) adjectives of plutonist and neptunist are used to indicate the followers of the two opposed doctrines.

However, his thesis led Hutton to two other propositions, mentioned above, which are truly innovative and have remained so to the present day. The problems of the uplifting of earth layers and of the origin of granite shall be investigated separately.

## Angular Unconformities

We owe to Hutton the discovery of what is known today as angular unconformity, by means of which geologists still recognize and date orogenic movements. Let us imagine that a series of horizontal rock layers is compressed and folded. The deformed layers are immediately attacked by erosion, which levels them gradually. The folds are thus truncated by the topographic surface, as if they had been cut off. If the area is subsequently covered by the sea, the newly deposited sediments rest "unconformably" on their substratum; that is, the new layers form an angle with the ancient folded and truncated layers. In extreme cases, folded layers are vertical and layers before and after the tectonic movement are at a right angle (fig. 9.1).

Angular unconformities allow us to date uplifting if the age of the rock layers is known. The movement is considered to have occurred after the youngest folded layer and before the oldest horizontal layer.

To follow Hutton's discovery of angular unconformities, one must read his *Theory of the Earth*. Fortunately, the chapters were written without being revised so that the stages of the evolution of his thinking can be found. In 1785 Hutton first read a dissertation at the Royal Society of Edinburgh: *Concerning the System of the Earth, Its Duration and Stability.*[6] An abstract was published soon after, and the complete text appeared in a journal in 1788, entitled: "Theory of the Earth; or an Investigation of the Laws Observable in the Composition, Dissolution, and Restoration of Land upon the Globe."[7] Finally, in 1795, the Scottish physician published the complete version of

Figure 9.1.    Hutton's Diagram of an Angular Unconformity (From Theory of the Earth, 1795). The lower vertical layers are the "roots" of ancient folds, which were partly destroyed by erosion. The peneplained surface was covered by the sea, which deposited the upper horizontal layers.

Theory of the Earth, with Proofs and Illustrations, in which the dissertation forms the first chapter.[8]

However, Hutton's first chapter and the three following ones in the first edition do not mention any unconformity. This shows that he added the idea later on. According to one of his biographers, Hutton supposedly wrote the main parts of his theory as early as 1760, and he understood the igneous origin of basalt before Desmarest.[9] However, so long as he lacked the concept of angular unconformity, the work remained unfinished.

## Predictive Theory

In chapter 5 of Theory of the Earth, the author wondered about portions of the earth that had been several times covered by water and then exposed to the air. He mentioned that some naturalists, such as J.-A. Deluc, had seen horizontal limestones resting upon folded schists—without, however, recognizing the significance of the situation. Hutton understood that these schists had been cut off by erosion, that is, exposed to weathering agents, before limestones were deposited on top. He *predicted*, therefore, the existence of angular unconformities.

In chapter 6, he stated his first observations on that subject. He at first saw layers inclined at an angle of 45° in opposite directions, thus representing the two legs of the Greek letter lambda. Then he observed horizontal layers resting upon vertical ones (see fig. 9.1).

This latter arrangement presented a problem if one followed Hutton's concept, because he stated that uplifting of layers also caused their tilting. If layers remained horizontal, they must consequently have been uplifted without any disturbance. But who can prove that they were uplifted? J.-A. Deluc, who had seen angular unconformities before Hutton, stated that if horizontal limestones had been uplifted, as had the schists on which they rest, they would have been "broken and disturbed in the same fashion."[10] This was common sense.

Fortunately, Hutton was not influenced by such commonsense arguments. After having predicted the observation in chapter 5, he went looking for these "junctions" or "contacts," which he then described in chapter 6 as observations on the island of Arran, at Jedburgh and Siccar Point.

What remained to be done was to date these superposed layers. Like his contemporaries, Hutton talked about primary and secondary layers. However, their age was determined by the angular unconformity, so that the folded strata were primary and the others secondary. Today, we know that it is not possible to date the angular unconformity itself, although it can give the relative age of the layers of rock formations. Thus, angular unconformity acts as an archive because it has the double function of revealing and dating orogenic events. Angular unconformities thus replaced, in a certain way, the lithological archives of Werner (and other neptunists), who associated age with the nature of layers.

Hutton believed that angular unconformities have the same age wherever they occur. Like neptunists, who believed that granite or gneiss were of the same age everywhere, plutonists had to accept the idea that the upheaval that had created continents was an event that had occurred at the same time over the entire globe. The history of mountain building could not be done on a regional basis.

Because Hutton was searching for a model of earth processes, he established a cyclic theory, where the same phenomena recur indefinitely. The concept led him to write the famous phrase, which seemed to reject the Creation and led to accusations of impiety: "We find no vestige of a beginning, no prospect of an end."[11] However, cyclicity destroys most traces of former cycles. We notice here again that cyclic theories are not suitable for historical purposes. In particular, as was the case for Hutton, cyclic theories based on the existence of deep erosion and melting by heat of materials depend on processes that tend to erase archives.

In fact, in his explanation of how land actually became habitable, he wrote that not two, but three successive worlds had to follow each other:

> If the earth on which we live began to appear in the ocean at the time when the last began to be resolved, it could not be from the materials of the continent immediately preceding this which we examine, that the present earth had been constructed; for the bottom of the ocean must have been filled with materials before land could be made to appear above its surface. . . . The world which we inhabit is composed of the materials, not of the earth which was the immediate predecessor of the present, but of the earth which, in ascending from the present, we consider as the third, and which had preceded the land that was above the surface of the sea, while our present land was yet beneath the water of the ocean.[12]

## The Igneous Origin of Granite

The theory of the igneous origin of granite is the third aspect of Hutton's plutonistic theory. It is perhaps the most daring element considering that penetrating minds such as Werner, Dolomieu, and Saussure still believed in the aqueous origin of granite as the first deposit at the bottom of the universal ocean. Again, the *Dissertation* of 1785 mentioned nothing; but in the summer of the same year, Hutton visited Glen Tilt where he observed that granite—obviously in a liquid state—had intruded rocks. He repeated his observations the following summers. In 1787, on the island of Arran, he also discovered angular unconformities.[13]

Hutton was accompanied by two students. One was John Playfair, who, as mentioned above, popularized the Huttonian theory in his 1802 *Illustrations of the Huttonian Theory of the Earth*.[14] The other was James Hall (1761–1832), known for his essays in experimental geology, one of which referred specifically to the question of intrusive granite.

Granitic intrusions drive back rock layers, which undergo important lateral stresses. Hall tried to show this in a model. He piled sheets of cloth horizontally, one above the other. He placed vertical boards on both sides and then pushed them slowly together so as to simulate lateral forces. Weights placed on top of the pile of cloth represented the overlying sedimentary cover. Thus, Hall obtained folded sheets corresponding to the large undulations he had observed in the field with Hutton and Playfair.[15]

Together with his two students, Hutton succeeded in presenting an explanation of all the orogenic phenomena: rising molten granite compresses, folds, and uplifts layers deposited at the bottom of the sea and makes them emerge as mountain chains, which later undergo the effects of erosion. But his theory and the observations on which it was based—angular unconformities on the one hand, and granitic intrusions on the other—were known by his contemporaries only through a short article published in 1794.[16] The other new findings were meant to form volume 3 of the *Theory of the Earth*. However, they remained in manuscript for a century, until 1899, when Archibald Geikie edited and published volume 3.[17]

There is no doubt that these new ideas surprised neptunists, who were busy classifying the layers of the earth but were little interested in such an ahistorical approach. However, among the neptunists were geologists who had also observed lateral compressions of layers and made an attempt to explain them.

## Saussure and Dolomieu

In his exploration of the Alps, Saussure had seen "layers forming a 'C' or an 'S'"; that is, recumbent folds with a nearly horizontal axis. He explained these folds as "a refoulement [horizontal **thrust**] which has folded" the left and right parts of a layer "one above the other," much like a pancake folded in two. What he believed to be the clue was the presence of an "empty space" at the site where the folded part had been before the movement.[18] Strangely enough, he stopped there, at least in his *Voyage dans les Alpes*. He had broken the Wernerian heritage, but seemed incapable of proceeding any further. Carozzi's study of his manuscripts shows, however, that Saussure actually had considered large-scale—even global—horizontal thrusting, and that he realized that this meant large-scale shortening of the crust and the underlying sliding planes. These ideas came close to the concept of contraction of the crust due to the cooling of the earth (chapter 12).[19]

Déodat de Dolomieu had also observed large-scale folding in mountains. He said that the structure of the Alps brings to mind "a shock that, hitting obliquely against the consolidated crust of our Earth, compressed this crust, broke it with violence, displaced and uplifted layers, and forced some of them to prop up or to support each other in a standing position, as is the case of Mont Blanc, while others fell after the shock and were thrusted over underlying masses . . . such as the rocks which form Monte Rosa."[20]

However, the larger the scale of the phenomenon, the less reasonable the solution. For instance, Dolomieu said that he had "a weak spot" for Whiston's system, meaning the explanation given in 1696 by William Whiston, disciple of Newton, in *A New Theory of the Earth* for the formation of mountains by the action of a comet.[21] An "extraterrestrial shock," said Dolomieu, that broke "the shell of the Earth" might well have produced the described effects.[22]

Saussure and Dolomieu lagged behind Hutton in their search for an explanation of compression (the effects of which they observed with as much precision as Hutton); whereas Henri Gautier (1660–1731), a civil engineering inspector in the Languedoc, France, had expressed quite early some revolutionary ideas with respect to tectonics.

## Gautier

Gautier's contemporaries knew his ideas well—Bourguet refuted him at length, though without naming him, and Maillet devoted twenty pages in his *Telliamed* to the same purpose—but he was completely forgotten later on. It took François Ellenberger's recent works to restore this author to the position he deserved.[23]

Gautier was aware of deformation of rock layers in mountains. In 1721 he wrote in his *Nouvelles Conjectures sur le Globe de la Terre* that, if sedimentary layers are comparable to "a recently constructed building," rock layers in mountains are similar to "another building which was knocked down, in which brick layers are overturned, appear upside down, thrown sidewise or some other way."[24]

Like Dolomieu and Saussure, Gautier accepted lateral movements to explain this deformation. The segments of the earth's crust "become superposed and form mountains . . . much like ice rafts floating on a river, which, when they encounter obstacles, override each other and form mountains of ice."[25]

But if we want to count Gautier as a forerunner of Hutton we must stop right there because in other respects he was very much an author of his own time. Indeed, he was even pre-Newtonian in his use of Cartesian physics, and he proposed a very strange representation of the structure of the planet, which probably led his readers astray. He believed that the earth was hollow, or rather filled with an "air much more subtle" than the atmosphere, so that the earth resembled a barrel "emptied of its wine." The crust of this globe was very thin, only 5,390 toises (about 10 kilometers). Even more bizarre was Gautier's belief that the interior and exterior surfaces of the crust were symmetrical; that is, seas and mountains existed inside the planet as well as on its surface (Gautier illustrated this idea with a sketch). He said that the reasons for this symmetric structure lay in two opposing forces: gravity and a "central force" caused by the rotation of the earth. The two forces canceled each other out in the middle of the crust. Thus, the central force dominated underneath and produced the same effects as gravity, but in the opposite direction.

The Huttonian theory is, of course, much closer to modern theories than Gautier's. For instance, Hutton's idea of a subterranean fire, going back to Descartes' central fire, is still popular today. Erupting volcanoes, even to people who have only seen them on television or film, suggest that the interior of the earth consists of molten matter, even if there is no proof that this "fire" reaches all the way to the center of the earth.

## Needham and the Steam Engine

Like Hutton, Dolomieu believed that below the earth's crust existed a subterranean fire that ejected basaltic lavas.[26] In 1769, before Hutton and Dolomieu, John Turberville Needham had proposed the same theory in the course of experiments he did with Buffon to prove spontaneous generation of living creatures. The concept of spontaneous generation had been seriously contested by Lazzaro Spallanzani. To a French translation of Spallanzani's essay, *Nouvelles Recherches sur les découvertes microscopiques* (New research on microscopic discoveries), Needham added a lengthy work, which translates in English to *Physical and Metaphysical Essay on the Nature of Religion, with a New Theory of the Earth, and Measurements of the Elevation of the Alps).*[27] Needham explained, in particular, that "a central fire which reached up to the surface of the Earth" produced the "internal expansive forces" that lift the earth's crust and produce plains when the terrestrial masses resist these forces, mountains (hollow inside) when they resist partially and become uplifted, or volcanoes when they (terrestrial masses) break.[28]

Needham's most original idea on mountain building was the analogy he saw with a "machine" using force "either produced by steam, or by extremely thin and dry air."[29] This model is interesting because it comes close to the image of the steam engine. Unlike Hutton, Needham could not have gotten the idea directly from Watt, because Watt only made his first (failed) experiments in 1769, the year Needham's book appeared. Nevertheless, the idea was very much in the air: as early as 1705, Thomas Newcomen had built a machine using the force of steam.

The invention of the steam engine was important not just because it domesticated fire and made it useful for humans, but also because it provided a model for the uplifting of mountains. Until that time, naturalists described the action of fire by comparing it to cannon powder; that is, an explosion that ejects matter by pulverizing it. This image described perhaps the chaos among volcanic debris but did not explain a slow uplifting of mountains. The steam engine thus presented a stimulus for further research by providing a new and better model to explain mountain building.

But how was this internal fire maintained? To run a steam engine, a source of heat is needed. Earlier, Descartes had described the earth as a cold star that retained in its center the matter of the first element. Leibniz and Buffon had adopted this theory, saying, however, that their star had cooled down to the core. J. J. Dortous de Mairan had searched for proof of a permanent internal heat source in the earth's core, at first in 1719, and then in 1765 (thus at the same time as Needham). However, he did not draw any conclusions in regard to mountain building.[30] Moro, who did assert the uplifting of lands, had sketched a world with a central fire, but only in a hypothetical fashion; he explained that uplifting did not create an empty space because a fluid earth fills in the space caused by the deformation of the crust.

Hutton, pressed by his adversaries, had to find a solution. The solution he came up with was a simple one and not at all new: combustion of coal. Most eighteenth-century naturalists had given this interpretation for the origin of volcanic fires. Nevertheless, Hutton made it a permanent cause, saying that each cycle forms new continents, which produce new forests; their destruction in turn forms new layers of coal.

## Werner and Hutton

To conclude this chapter, the merits of Werner and Hutton and their respective roles in the establishment of geological science are contrasted and compared. One should remember that Hutton's "plutonistic" explanation of diagenesis of rocks (namely, induration) was erroneous. But he understood the origin of granite better than Werner did. His student, James Hall, succeeded in crystallizing calcium carbonate powder into limestone. In a sealed metal tube, calcium carbonate was heated under pressure to the melting point of silver in 1805.[31] These were the first laboratory tests in experimental metamorphism. Yet it took another century and a half to produce granite by fusion (experimental anatexis, that is, the process by which igneous rocks remelt into magma). Experiments by J. Wyart and G. Sabatier in France and by H.G.F. Winkler in Germany in the 1960s demonstrated that sediments, at a pressure of 2,000 bars and a temperature of 800° C, become a liquid with the composition of granite. The formation of granite is therefore a phenomenon that has always existed and which occurs at great depth. It is not an ancient and unique event that occurred only in the "primitive" epoch of the history of the earth.

Thanks to James Hall's laboratory demonstration, the igneous origin of granite was accepted rather quickly, and geology became Huttonian. Geology kept nothing from the Wernerian neptunism except the historical preoccupation that Hutton's theory lacked. But without that sense of geological history, Hutton's cyclicity would not have been of any use.

Hutton triumphed also in matters of tectonics. Werner was not interested in deformation of strata. His system forbade it by its claim that rock layers had kept their order of deposition. Hutton, on the other hand, proposed a key for the recognition of uplifting. In that respect, he complemented Steno's tectonic principle of deformation of beds after deposition (see chapter 5) and provided, 130 years after Steno, the second milestone of modern geology. The methodology of both Hutton and Steno became an integral part of science and allowed the geognostic chronology to be completely redone and a new, definitive understanding of earth's history to be established.

Because Hutton was not interested in the history of the earth, he was to neptunists what Newton was to Cartesians: he abandoned history for the benefits of an operational law. However, with that law, he provided the key that allowed his successors to read history.

## Repeated Uplifts

At the same time, Hutton proposed the idea the neptunists lacked: that the history of the earth is based on repeated phases of uplifting of mountains. This idea was not new; the cycles of the Stoics represented the first version of this concept. However, the Stoics were long forgotten.

Hutton's era may be considered a kind of return to the ideas of the Stoics. Indeed, in the middle of the eighteenth century, Nicolas-Antoine Boulanger visualized a cyclic system with periodical restorations of the earth (it was just as unhistorical as Hutton's because each cycle erased the traces of former cycles). Gautier, before Boulanger, also proposed a geology punctuated by orogenic cycles. Finally, two famous contemporaries of Hutton, each in his own way, presented ideas of cyclicity.

The first was Jean Baptiste Lamarck (1744–1829), who is best known for his ideas on evolution. But he also tackled geology. In his book *Hydrogeology*, published in 1802, he argued that the axis of the poles changes imperceptibly together with the "equatorial bulge"—the swelling of the earth around the equator.[32] That swelling is greater for continents than for the ocean. Therefore, the highest lands are at the equator because the ellipsoidal shape of the earth protrudes more than the ocean does. Hence, land dominates over the ocean at the equator. Because the axis of the poles rotates about the earth in 9 million centuries, the lands and the seas undergo a cycle that brings to mind Buridan's long cycle (chapter 2). Boulanger also mentioned changes of the axis of the poles and of the shape of the terrestrial spheroid, but he claimed these changes were more violent in character.

The second author was Jean-André Deluc, a geologist from Geneva who lived in England for a long time and was one of Hutton's adversaries. He believed that mountain building required "revolution upon revolution." However, these catastrophes were not uplifts but collapses caused by internal cavities according to the old ideas of Leibniz and Buffon. Deluc was appalled by the disorder of layers, which looked like "masures" or "buildings in a state of ruins."[33] Whatever angular unconformities he might have seen persuaded him that rock formations were not uplifted.

Nevertheless, it would be unfair to laugh at Deluc's archaic tectonics or his concerns as a scrupulous Christian about reconciling

Scriptures and geological observation, worked out in particular in his 1798 *Lettres sur l'histoire physique de la terre, adressées à M. le professeur Blumenbach, renfermant de nouvelles preuves géologiques et historiques de la mission divine de Moyse* (Letters on the physical history of the earth, sent to Professor Blumenbach, including new geological and historical proofs of the Divine mission of Moses). Indeed, when he explained that *present* mountains were caused by repeated orogenies, he introduced the notion of tectonic phases into the history of the earth. He thought that mountains were not destroyed from one stage to the other; on the contrary, they became progressively higher by the cumulative results of "revolutions." If he did not understand the importance of angular unconformities, at least he noticed the repetitive character of mountain building. More important, he was one of the first to understand the use of fossils in stratigraphy, as we shall see in the next chapter.

# Chapter 10

# The Use of Fossils

## Deluc

Jean-André Deluc (1727–1817) and his younger brother, Guillaume-Antoine (1729–1812), had always been interested in fossils. Born in Geneva, Switzerland, they were pioneer mountain climbers and collected many "marine bodies" high in the Alps. Jean-André published a series of works between 1778 and 1809, mostly in the form of letters, which allow us to follow his thinking. He was already fifty years old when he wrote his first work. Hutton, too, who was one year older than Deluc, was unknown to contemporary naturalists until he was quite old.

Deluc's first *Lettres physiques et morales* (Letters on natural history and ethics) were addressed to the queen of England. They referred to Switzerland and southern France. The next set of letters, under a very similar title, described his travels through various regions of Europe.[1]

Between 1790 and 1794, Deluc sent a series of letters to J.-C. de Lamétherie, editor of the journal *Observations sur la physique*, where these letters were published. Thereafter, Deluc wrote letters to Blumenbach (see chapter 9), which were published in an English journal in 1793 and 1794. Finally, his *Traité élémentaire de géologie* (An elementary treatise on geology) appeared in 1809 without providing any new developments in comparison to his earlier works.[2]

It was in his letters to Lamétherie that Deluc was most concerned about establishing an order of deposition of layers. He had observed on the coast of southern England and of the Isle of Wight a superposition of "three classes of layers" and wrote, "layers of clay occur underneath limestone layers which are overlain by chalk."[3] He determined their relative age by using the principle of superposition, proposed by Steno (chapter 5) and rediscovered by Soulavie and Werner. Deluc's study is important because of his emphasis on the fact that these layers differed not only in their lithological nature, but also in the fauna they contained. For instance, the clay contained "flat oysters," and in some places, ammonites. The latter were also found in the limestone, but were missing in the **Chalk,** which was therefore "younger than the revolution which destroyed ammonites together with other shellfish." But the Chalk contained sea urchins with stubby spines and various other fossils "which are not found in any sea" at present; whereas "univalves" (gastropods), which are currently very abundant in the English Channel, do not occur in the Chalk.[4]

## Index Fossils

Deluc entered here a field of research that was not entirely new, since Palissy, and later Hooke, had observed that some species were "lost," in contrast to Leibniz, who preferred transformation of species as a solution to the problem.

But it was not enough to observe that layers contain different fossils and merely guess that these fossils could be used for chronology. Indeed, Rouelle and his students continued to think—in the mid-eighteenth century—that layers were more or less the same age and that faunal changes indicated variations in depth and climate. Recall, too, that Martin Lister, at the end of the seventeenth century, had used these variations to propose that fossils were formed in the same manner as stones.

Two steps can thus be distinguished in the discovery of the stratigraphic role of fossils. The first was to observe faunal variations from one layer to the other. This had been accepted at the time of Rouelle, and predicted at the end of the preceding century. The second step was to explain this variation by a succession of forms through time. Deluc was aware of this.

Before Deluc, Soulavie had recognized three "ages" of rock units based on fossils—depending on whether they contained only forms without modern analogues (ammonites, **belemnites**), or a mixture of species that had disappeared and modern species, or, finally, only organisms that still exist today (sea urchins, oysters, gastropods, and so on).[5] Similarly, Arduino had noticed in 1783 that fossils found in tertiary hills were lacking in secondary mountains and vice versa. However, his influence was limited because his most innovative ideas remained in manuscript form.[6]

## Synchronous Histories

For biostratigraphy to be born, a third step had to be taken: fossils had to become the means to date layers. Deluc began by establishing a superposition of layers, that is, their relative age, and then noting the fauna included in each layer. He defined the age of layers, as well as that of fossil species, according to the principle of superposition. Based on that principle, Deluc started to write "two parallel histories: that of rock layers and that of organized beings."[7]

Fossils were thus not being used to determine the age of layers, but on the contrary, the age of deposits was used to date ancient fauna. Nevertheless, it was a start toward testing the new "chronometer." The Wernerian school had actually done the same thing. The order of superposition of rock units had provided a geognostic scale that could be used later in areas where the order was hidden (it might even have served to date overturned layers, had Werner recognized them). More importantly, the geognostic scale was used to put thin local sections—so often encountered in the field—into the universal column. Consequently, if the faunal variations Deluc established in England had the same universal character, one could tell, for instance, if the cliffs along the coasts of France were of the same age as those of England.

Two problems—which geognosy had already encountered—arose immediately for Deluc. First, he needed to discover a global process that changed faunas in order for faunas of the same age in different sites to be similar, or the changed species at least had to be capable of moving rather fast in order to spread rapidly into all oceans. Second, he needed to discover the law regulating the processes of variation.

## A Transformist

Deluc proposed an explanation that accounted for both deformation and variation of composition in layers. He said that the interior of the earth had compacted, causing layers of land to collapse. As a result, "expansive fluids" escaped, which modified the composition of the ocean (that is, later deposits formed during precipitation from the marine "liquid") and led to the disappearance of living organisms. This disappearance was twofold: some species were destroyed, others underwent "great changes" or transformations.[8]

Deluc's "transformism" followed the one proposed by Leibniz a century before. Somehow, the concept did not seem incompatible with any loyalty to Scriptures. Although transformism was theologically offensive when it gave nature the power to modify the divine work of Creation, it was nevertheless a lesser evil because it avoided a more radical solution: spontaneous generation of species. Deluc explained that he was refuting spontaneous generation and therefore adopting the idea of transformism to account for faunal succession. However, when Soulavie said in the first draft of the first volume of his Histoire naturelle de la France méridionale that "recent families which do not exist in ancient marbles . . . descend from primordial families," the idea of a "metamorphosis of several species" was immediately condemned by the Church.[9] Soulavie was forced to change his text, purging it of such heretical statements before publication. What shocked the faithful even more a few years later in Lamarckian transformism was perhaps the assumption of the spontaneous generation of very simple organisms, and of course also his statement that humans descended from animals.

The next problem after attempting the concept of evolution was the search for a law regulating this process. It should be remembered that the Wernerians had proposed such a law to explain the superposition of layers. As a result, they believed that if a law of evolution could be established, it would no longer be necessary to observe all superpositions of layers to build a geognostic column. If, for instance, rocks decreased in crystallinity upward in a stratigraphic column, or if, as Leopold von Buch said—before becoming a dissident and joining the plutonists—their nature was "farther and farther removed from alkalis," one could define the stratigraphic position of any rock unit through a simple chemical analysis without having to study, hammer in hand, its relationship to nearby rock units.[10]

A similar law is possible in paleontology. Soulavie based his dating on the relative percentage of present-day organisms in rocks. This method has its value. Charles Lyell, for example, following the statistical studies of G. P. Deshayes,[11] divided the Tertiary era into four periods: Eocene, Miocene, Old Pliocene, and New Pliocene.[12]

Lyell's divisions, as well as the modern terms of **Paleocene, Eocene, Oligocene, Miocene,** and **Pliocene** for the Cenozoic era, and the term Pleistocene for the **Quaternary** era, were established on the basis of relative percentage of present-day species from one period to the other.

However, Soulavie had in mind another type of law when he said that nature changes from a simple to a complex state. This was not a new concept. We could even say that it is as old as natural history itself since Aristotle used it. Nevertheless, the concept changed its meaning with respect to transformism, or the theory of successive creations of species, and described the order of appearance of organisms and not simply a progression through theoretical successive stages.

Arduino was of the same opinion when he wrote in a letter to Antonio Vallisnieri, Jr., in 1760 that lower strata of secondary mountains contain organisms of lesser perfection.[13] This means that these authors were able to begin the third stage of the chronological use of fossils by providing rules on how to establish the age of rocks based simply on the observation of organic remains.

## Cuvier and Brongniart

The decisive moment of the establishment of this last stage was the publication of a famous memoir by Georges Cuvier and Alexandre Brongniart, *Essai sur la géographie minéralogique des environs de Paris* (Essay on mineralogical geography of the surroundings of Paris), published first in 1808, in an article of fewer than forty pages, then expanded to almost three hundred pages in its final form in 1811.

The two authors already had an excellent reputation. Cuvier, born at Montbéliard, on German territory, studied at the famous Caroline University (Hohe Karlsschule) at Stuttgart. He spent some time in Normandy, where he was tutor to the son of a local nobleman, and then went to Paris in 1795. In the meantime, the territory of Montbéliard was annexed to the French republic, and Cuvier became a French citizen. He found a post as assistant professor of animal anatomy at the Museum of Natural History, founded by the National Convention during the French Revolution. He became full professor in 1802 and renamed his chair comparative anatomy. Meanwhile, in 1800 he had succeeded Daubenton, Buffon's former co-worker, as the professor responsible for teaching natural history at the Collège de France. Without taking into account the administrative functions he accepted later on, he already held two very important university positions.

Alexandre Brongniart is particularly famous for having been director of the Sèvres porcelain factory for forty-seven years (from 1800 until his death). He was a great administrator, created workshops for glass and enamel painting, and founded a museum of ceramics. In 1811, the year of the publication of the memoir, he inaugurated the teaching of mineralogy at the newly founded Faculty of Sciences at Paris. Later on, he succeeded the mineralogist René Just Haüy at the museum.

In their memoir the two naturalists were interested in the Chalk, the last layer of the Mesozoic, and the overlying Tertiary rock units in the Paris Basin. The authors noticed that the superposed rock layers differed in their nature, in their thickness, and "especially in the fossils they contained."[14] The expression "especially" shows the importance Cuvier and Brongniart gave to fossils, although fossils were not, as yet, understood in terms of evolution and environment as in modern **biostratigraphy.**

Soon the study of fossils produced interesting results. In 1821 Brongniart wrote in an article that the Chalk contains similar fauna in France and Poland. Referring to a rock unit in the Mountains of Fis (massif of the Buet), which had been classified by Wernerians among transition rocks (today known as Paleozoic), he showed that its fossils were those of the Chalk (end of the Mesozoic). He asked the fundamental question, "When petrographic and paleontologic criteria differ, which one is more reliable?"[15]

This problem was, of course, everyone's concern. Although the traditional method still had some followers, it was now asked why rock units of a same epoch should be identical when present-day deposits vary from place to place: sand in one place, clay in another, and so on. The Wernerian theory's proposal of a universal crystallization had no modern equivalent, and the theory lost its prestige among contemporaries of Cuvier and Brongniart.

The paleontological method also had its problems, because fauna is not identical everywhere in the world. To justify stratigraphy, the earth should really have had "the same climatic conditions" during each epoch.[16] This meant that the past was in fact quite different from the present. At the heart of these discussions lies the problem of the relationship between the present and the past, namely, the problem of uniformitarianism (see chapter 11). Cuvier, however, was not concerned with uniformitarianism. Instead, he proposed to explain faunal variation by "revolutions."

## Destruction and Creation

In his foreword to his four-volume *Ossements fossiles* (Fossil bones), published in 1812, Cuvier approached the subject of revolutions. Republished in a separate volume, the foreword became the famous *Discours sur les révolutions de la surface du globe* (A discourse on the revolutions of the surface of the globe), reprinted several times during the century.[17]

Cuvier began with Brongniart's fundamental demonstration on alternating marine and freshwater deposits.[18] For people who were used to reasoning in terms of one retreat of the ocean, this demonstration alone stirred up a small revolution. However, Cuvier shocked them even more when he tried to explain the phenomenon, whereas Brongniart had remained cautious. Cuvier said that sudden revolutions had either dried up areas once covered with water or flooded dry regions. Impressed by the study of mammoths, with which he had started his career in Paris, he was convinced that only a sudden catastrophe could have trapped animals in ice.

Cuvier wrote that violent phenomena may destroy fauna that

## The Short Life of Species

To be sure that similar animal species have lived at the same time on every continent, it would suffice to know that each species lived only a limited time. If species X lived between time T and time T', one could be sure that the rock unit that contained the fossil X was deposited in the interval T–T'. This is what happens according to evolutionary thought. The shorter the interval T–T', the more precise the dating of a species.

However, in the beginning of the nineteenth century, it was not necessary to accept transformism to use this argument. Species X may appear at time T after transformation of species W; it may also appear at T by creation or by spontaneous generation. It may, furthermore, disappear by destruction as well as by evolution (evolution by natural selection does not, of course, deny destruction of species).

## A Gigantic Freezer

The discovery of frozen animals in Siberia has much contributed to the theory of revolutions on the earth. Cuvier said that the last of these catastrophes

> left in the northern countries the remains of large quadrupeds frozen in ice and thus preserved until today with their skin, their hair, and their flesh. Had they not been frozen immediately after death, putrefaction would have decomposed them. On the other hand, this eternal ice did not exist before in the areas where the animals were frozen because they could not have lived in such climates. Therefore, these animals were killed at the same moment as the countries, where they lived, were exposed to colder climates.*

In 1772, Pallas found on the banks of a tributary of the Lena River in Siberia the body of a rhinoceros. He wrote:

> The skin had preserved its external organization: one could see some short hair, even the eyelids were not completely decayed. In the socket of the brain, and here and there under the skin, I found some matter which was the residue of decayed parts of flesh. . . . Only at the time of the Deluge could this animal have been transported from the southern countries into these frozen areas of the north.†

Lyell reported that in 1803 the whole skeleton of an ice-locked mammoth was found farther north on the banks of the Lena River:

> The skin was covered first with black bristles, thicker than horsehair, from twelve to sixteen inches in length; secondly with hair

cannot live in new environments. However, how did the area become repopulated? Here, the author of *Discours* was cautious. He explained that if a narrow strip of land suddenly joined Australia to the Asian continent, land species would migrate and replace insular organisms destroyed by a former catastrophe. This was the only solution Cuvier ever gave.[19]

However, this solution implied that fauna new in one region had previously existed elsewhere. It did not explain the appearance of fauna that had not existed anywhere on the globe. Therefore, Cuvier's solution did not relate each fauna to an epoch and thus did not allow long-distance correlations. This is why Cuvier's successors

*(continued)*

of a reddish brown colour, about four inches long; and thirdly with wool of the same colour as the hair, about an inch in length. Of the fur, upwards to thirty pounds' weight were gathered from the wet sandbank. . . . The species may, as Cuvier observed, have been fitted by nature to withstand the vicissitudes of a northern climate.[‡]

It was therefore not necessary to imagine that Siberia had once been an equatorial climate as Buffon believed. Mammoths and rhinoceros with septate nostrils preserved in the ice of Siberia were animals adapted to the climate of cold countries.

---

[*] Georges Cuvier, *Discours sur les révolutions de la surface du globe et sur les changements qu'elles ont produits dans le règne animal* (Paris: Dufour et d'Ocagne, 1825), 16–17.

[†] Peter Simon Pallas, *Voyages du professeur Pallas dans plusieurs provinces de l'empire de Russie et dans l'Asie septentrionale,* trans. from the German, 2d ed., 8 vols. (Paris: Maradan, an II, [1793–1794]), 5:215–218.

[‡] Charles Lyell, *Principles of Geology, or the modern Changes of the Earth and its Inhabitants.* . . . 10th rev. ed., 2 vols. (London: John Murray, 1867–1868), 1:184. Lyell refers to Georges Cuvier's *"Ossements fossils,"* 4th ed. (n.p.: 1836). There is evidently a discrepancy between Cuvier's statement in regard to frozen animals in Siberia that "could not have lived in such climates" and Lyell's remark that "the species may, as Cuvier observed, have been fitted by nature to withstand the vicissitudes of a northern climate."

preferred to think about general destructions of species followed by new creations.

However, repeated creations presented a theological problem. They made it necessary to assume that God had tried several times to create animals. And because it was agreed that successive creations were increasingly perfect, God must have failed in his first trials. The question was therefore asked: If the present state of creation is not necessarily final, why should there not be a more perfect human being in the works?

The idea of multiple creations associated with the idea of progress had come up in earlier works. Cuvier distinguished several successive epochs—which, in fact, contradict his thesis of migrations and show that he was not content with it—but he remained cautious, surely to avoid any religious implications. According to Cuvier, the oldest rock units contained remains of invertebrates: mollusks, crustaceans, and perhaps fish. Thereafter came the oviparous quadrupeds (reptiles). Still later were found the remains of marine forms of mammals (cetaceans), shortly before terrestrial ones.[20]

Here we again have the two laws of pre-biostratigraphy. First, a universal mechanism was assumed to cause faunal replacement so that long-distance correlation of fauna becomes possible. Second, a law of succession of animal forms was sought to provide, so to speak, a clock for telling time so that time relationships would not have to be reconstructed case after case. If there was a progression of forms, the degree of complexity of species would be the direct measure of the age of rock units.

In 1832, at the time of Cuvier's death, biostratigraphy tended to replace the former chronology (**lithostratigraphy**), which was based on the nature of rocks. Brongniart, who was less theoretical than Cuvier but a better observer in this field, worked hard to show that the division of geologic time can be based on paleontology. I shall return to the remainder of his work in chapter 14. For the time being, I must go back to the actual beginning of biostratigraphy.

## William Smith

In 1816 the English land surveyor William Smith (1769–1834) published a small work, *Strata Identified by Organized Fossils*, in which he established a list of seventeen strata with their characteristic fauna for the rock sequence between the **Jurassic** and the **Tertiary**. The following year, he added ten more strata downwards until he reached, he said, the primordial granite and gneiss.[21]

Cuvier and Brongniart's work (1808) was published eight years before Smith's. But according to Smith, his manuscripts had circulated as early as 1799, and his student and disciple John Phillips described observations made in 1791. However, Smith's priority remains controversial because he does not seem to have understood the problem before Cuvier and Brongniart.[22] He made excellent observations, but theoretical explanations were not his forte.

Furthermore, it seems that Smith's ambition was detrimental to his work. He distinguished some twenty divisions in the entire geological record, at a scale that differed greatly from that of Brongniart and Cuvier, who merely tried to subdivide into units Tertiary rocks in the hills of the Paris Basin. By focusing closely on time and space, their work had a greater reliability than Smith's. Indeed, new knowledge often results from concentrated effort and a narrowing down of the subject investigated.

However, priority does not really matter in this general history of geology. What I am trying to show are new ideas and, if possible, the reason why they emerged, even though I might be wrong with respect to the possible forerunners, in particular when the new contribution was actually the work of several persons. This convergence of ideas shows that the idea was in the air and that several contributions occurred at the same time.

## Lavoisier

Perhaps the real forerunner of Brongniart and Cuvier was Lavoisier. It is common knowledge that Antoine-Laurent Lavoisier (1743–1794) was the founder of modern chemistry. But it is generally not known that the illustrious savant's first and perhaps last scientific concerns were for geology.

Lavoisier was a student of Buffon, Rouelle, and Guettard between 1763 and 1766. He was associated with Guettard in the preparation of the *Atlas minéralogique de la France* (Mineralogical atlas of France). For this purpose, he traveled in the area of Paris and, between 1767 and 1771, in Alsace, Lorraine, and Flanders. His work in chemistry kept him busy thereafter, and it was only much later that he started geologic fieldwork again, working in Brittany (1778) and in Burgundy (1787). In December 1788, he read at the Academy of Sciences a memoir, the title of which translates to "General Observations on the Modern Horizontal Layers Which Were Deposited by the Sea, and on the Conclusions That Can Be Drawn from Their Arrangement with Respect to the Age of the Earth."[23] He ended his memoir with the promise of a paper to come, a promise which he could not keep because he was a victim of the French Revolution.

Lavoisier's insight can be noticed in his first works. In 1767 he told Bertin, who was in charge of the Technical Ministry of Agriculture and Mining, that in the kind of mapping Guettard had asked him to do, it was important to indicate the sequence of rock units.[24] This fundamental remark shows the break separating Lavoisier from

Guettard and marks the beginning of a new idea. This is why I shall backtrack here for a moment.

In a memoir read at the Academy of Sciences in 1746, Guettard presented his new work (the title in English is "Mineralogical Map on the Nature and the Location of Formations Which Occur in France and England"),[25] perhaps in response to a wish that Fontenelle, secretary of the academy, had expressed twenty-five years before.[26] With great determination, Guettard had undertaken an ambitious enterprise: to cover all of France with large-scale maps with the aim of producing an *Atlas minéralogique* in 214 plates.

The enterprise was too ambitious. In the beginning Lavoisier was the only help Guettard had, and hence only forty-five sheets were published. Furthermore, the publication of the last maps by A. G. Monnet brought about a quarrel because Lavoisier believed that these maps were not ready for publication.

Guettard had a purely utilitarian goal in mind: he wanted to provide data on the location of stones suitable for making lime or for construction, on various mines, and so forth. He was therefore content with the juxtaposition of "strips" of rock units of different natures, whereas Lavoisier cared about the relative position of these layers, that is, their order of superposition. Lavoisier therefore added cross-sections in the margins of the maps of the *Atlas*. He was right to do so because stratigraphy is expressed better by vertical cross-sections than with maps. Lavoisier's memoir of 1788, which contained several geological cross-sections of the Paris Basin, appeared posthumously in the volume of 1789 and was published in 1793 in the works of the academy.

Cuvier and Brongniart must certainly have profited from Lavoisier's ideas. Nevertheless, they still deserve credit because their contemporaries had not seen Lavoisier's point. Cuvier and Brongniart's essay included not only a "geognostic map of the vicinity of Paris," at the scale of $1:200,000$, but also several cross-sections in color surveyed with a barometer.

Mapping had rarely been done before that time (we have mentioned a map by Soulavie; Desmarest, as well as several other authors in the years 1770 to 1780, might be added); but it became suddenly of great interest throughout Europe. Cuvier and Brongniart's map (1811) was in fact the first modern geological map of France, or of a European country.

Their map was followed in 1815 by William Smith's *A Delineation of the Strata of England and Wales, with Part of Scotland,* which included fifteen plates drawn at the scale of five miles per inch.[27] It was the first of such small-scale geological maps, preceding by a quarter of a century its French equivalent, *Carte géologique de la France au 1:500,000,* completed in 1841. Two young mining engineers, Armand Petit Dufrénoy (1792–1857) and Léonce Élie de Beaumont (1798–1894), were in charge of the map of France. They began their fieldwork in 1825 after a visit to England. Data coordination began in 1830. On November 30, 1835, a draft of the map was presented to the academy and published six years later.

As the main French supporters of catastrophism, Dufrénoy, and in particular Élie de Beaumont, were to play an important role in geology later on. Before analyzing their ideas, let us turn to the rival school, uniformitarianism, whose tenets were presented in 1830 in the first volume of *Principles of Geology* by Charles Lyell.

Chapter 11

# Uniformitarianism versus Catastrophism

## Principles of Geology

In December 1831, the young Charles Darwin (1809–1882) embarked as naturalist on the *Beagle's* voyage around the world. Among the books he carried with him was the first volume of *Principles of Geology*, published a year before by Charles Lyell (1797–1875). Although "the sagacious Henslow," Darwin's professor of botany, "had recommended the book to him with the reservation not to accept any of the ideas," he was fast under the spell of Lyell's ideas. After his first observation of geology, he "was convinced of Lyell's infinite superiority of ideas."[1]

These ideas are traditionally summarized by the word "uniformitarianism," or by the term "present-day causes"—both explain that the "present is the key to the past." In opposition to geologists of his time who saw traces of former catastrophes everywhere, the author of *Principles of Geology* attempted to explain "former changes of the Earth's surface, by reference to causes now in operation."[2]

To embrace such an idea, it was certainly necessary to become aware of the great duration of geologic time. This is exactly what Charles Lyell did during his travels, which brought him in 1828 to France and then to Italy.

Born in Kinnordy, Scotland, the future geologist learned to observe and collect insects in the countryside. In 1818, after the customary classical education, he attended lectures on mineralogy and geology given by William Buckland at Oxford. His interest in geology was aroused. In 1819 he was elected member of the Geological Society of London; from then on he maintained his interest in geology while continuing his studies and practice of law.

In 1823 he spent two months in Paris, where he met many French scientists, but it was not until 1828 that he went to see geology for himself. Accompanied by Roderick Impey Murchison, he traveled through Auvergne, where he became particularly interested in stratified freshwater deposits. Observing layers thinner than one millimeter, he attributed each layer to an annual deposit and calculated that the entire series of layers (230 meters) must represent hundreds of thousands of years of deposition.[3]

His *Principles of Geology* reaped immediate success. When volume 3 appeared in 1833, the author had already revised the first two volumes (published in 1830 and 1832, respectively). In 1834 he restructured the entire work into a new edition in four volumes.

In 1838 Lyell published *Elements of Geology*, applying uniformitarian principles to the history of the earth.[4] The two works were revised several times before his death. Although Lyell is still considered the father of geology in many textbooks, or at least the man who popularized Hutton's ideas, I shall compare below his main ideas with the work of his contemporaries.

## Constant Prévost

To start with, uniformitarianism was not Lyell's invention. Constant Prévost (1787–1856), whom Lyell had met in Paris in 1823 and who, together with Ami Boué (1794–1881), was the founder of the Geological Society of France, had mentioned the same ideas in opposing Cuvier's school. As early as 1821 Prévost observed "a mixture of marine and fluvial [i.e., from rivers] shells in the same layers" in the hills of the Paris Basin.[5] This fact shed doubts on the repeated marine invasions postulated by Cuvier and Brongniart. Prévost believed that marine shells found above the "calcaire grossier" (coarse limestone) might be merely "reworked" fossils; that is, remains from that formation which had been exhumed by rivers and introduced into later deposits.[6]

Modern geologists know how easily the phenomenon of reworking can mislead them in the dating of a rock layer. Careful examination is necessary to find traces of abrasion, of having been transported some distances, or of a matrix from an earlier deposit. But in 1820, such detailed observations were not conceivable, and Prévost's remarks show fair caution.

Without even considering reworking processes, river mouths and estuaries are also known to show mixtures of marine and freshwater faunas similar to those Prévost thought he had found in the Cenozoic layers in the vicinity of Paris. He thus doubted the theory that present lands "have been several times covered by the sea" and preferred the old concept of the universal retreat of the primitive ocean.[7]

Prévost was thus a uniformitarian when he looked for present-day equivalents to what he observed in rock units and also when he refuted "revolutions on the Earth." Nevertheless, he did this in a curious fashion, taking advantage of some mixtures of marine and fluvial species in order to deny the alternate superposition of layers formed in the sea and in freshwater. He spent his entire career working on a "theory on tributaries," according to which two parallel types of deposits had existed through geologic times: a marine type formed essentially by limestones and a fluvial type consisting of coarse rocks such as sandstones. Superposition of the two types occurred only when they met, namely, at the mouths of rivers. Hence, during the Cretaceous, for instance, deposits of green sands and clay, called "Gault" (which occur in reality underneath the Chalk), were interpreted by Prévost as a formation of the same age as the Chalk. It happened to be underlying the Chalk locally only where estuaries deposit sand and argillaceous mud.[8] Prévost's approach shows one of the pitfalls of uniformitarianism. Based on present-day causes, he tried to prove that **transgressions** of the sea had not occurred, which is, in fact, a step backward.

## Uniformitarianism

To understand uniformitarianism, or the principle of present-day causes, which rules contemporary geology, we should perhaps separate two ideas associated with that concept: continuity and equilibrium. When William Whewell (1794–1866) coined the word *uniformitarianism*, he meant to describe Lyell's doctrine; catastrophism was, according to him, the opposite theory. Thus, Whewell emphasized only the degree of violence in phenomena invoked in the two theories.[9] However, instead of uniformitarianism and catastrophism, I believe that continuity and discontinuity better describe the two opposing theories.

## Continuity

Uniformitarians certainly fought for the concept of continuity in the earth's history. This is shown, for instance, in a debate between Prévost and Dufrénoy at the Geological Society of France in 1833. Dufrénoy wanted Prévost to admit that it was possible to separate the Tertiary era into three periods, each one with its typical fauna. Prévost gave in, on the condition that Dufrénoy recognize intermediate stages.[10] In fact, continuity in transitional stages meant refutation of violent catastrophes. Indeed, if transitions were slow, causes could be found in present-day processes.

Something similar happened to Darwin when he observed coastal uplifting in South America. Noticing that the shores of Chile had been elevated "imperceptibly" after the earthquake of 1822, he concluded that "earthquakes, volcanic eruptions, and sudden uplifting of the Pacific coast must be considered irregularities of a much larger phenomenon."[11] In other words, even discontinuities observed in *present-day* processes were suspected of not expressing the essence of geologic phenomena. One might almost talk of super-uniformitarianism!

## Equilibrium

Continuity was only one aspect of the theory of uniformitarianism. Lyell believed that both variation and its effects were uniform in intensity. More precisely, according to him the world was in a relatively stable state. Martin Rudwick qualified this concept as the "steady state model."[12] In other words, we could say that Lyell was a believer in both "steady state" and uniformitarianism.

The opposite doctrine was the directionalism of catastrophists, who assumed that the world evolved in a specific direction. We could perhaps use the simpler term of evolution if it were not apt to cause a possible confusion with Darwin's evolutionary biology.

In the first edition of *Principles of Geology*, Lyell believed in the stability of the living world. Of course, some species disappeared, while others were created; but species were somehow replaced by very similar forms, which were used for dating but did not change the general equilibrium.[13] And all this was done with discretion for the sake of continuity.

Prévost adopted very similar views. He was forced to recognize that, at the level of species, "ancient organisms of all classes are different from present ones." But he added immediately that "the physical structures of ancient organisms are not essentially different from those of present organisms."[14] Therefore, the latter "could have adapted to the environment at the earth's surface when the Late Paleozoic rock units were deposited."[15]

## For or Against Physical and Biological Stability

Uniformitarians wanted to prove the stability of the physical world. Across many centuries, their arguments joined those of Aristotle. They believed in variation, but also in processes that compensated for each other so that, on the whole, nothing changed except for a few minor details. In 1830 there was no reliable way to reconstruct the early history of the physical world, but faunas seemed to offer the best tools. It was well known that organisms had changed over time, and stratigraphy used fossils with increasing success to date rock units. However, to demonstrate an approximate stability of the physical world, it was sufficient to show that biological organizations had remained quite similar.

Who supported the opposite view? Catastrophists? Yes, certainly, because Cuvier stated that variations of the "liquid" "caused" faunal changes.[16] But they were not the only ones to postulate variations in the physical world. In 1835, Étienne Geoffroy Saint-Hilaire (1772–1844) claimed that if ancient fossil species returned to earth they would die because "the present environment would no longer provide required conditions for their respiration."[17] Geoffroy was a transformist. A disciple of Lamarck, he was the most respected supporter of the transformist doctrine of the time. The controversy over the unity of biological organization, which pitted him against Cuvier at the Academy of Sciences in Paris in 1830, would have distressed the old Goethe, who had studied the same subject.

## The Novelty of Catastrophism

In 1838 it was believed, as Jean-André Deluc had said forty years earlier, that the physical and the biological world proceeded according to a "synchronous" history.[18] When Deluc began his battle against uniformitarianism, he coined a new term to describe what appeared to him the traditional methodology in geology. Although, as we mentioned in chapter 10, he was the first to use the expression "present-day causes," the notion was obviously so accepted by everyone that no explanation was necessary. He rejected the uniformitarian methodology because he had observed that the earth had changed, both biologically and physically, and that the physical changes had caused the biological.

Cuvier (borrowing from Deluc without giving him credit) used the same expression to stress his disapproval of his predecessors, who "had believed for a long time that they could explain present-day causes by former revolutions. . . . However, the chain of events has been broken."[19] Cuvier thus emphasized both discontinuity and directionalism. According to him, ancient causes had been more powerful and they had acted upon a different nature, in particular, on a "liquid" (ocean) of different composition. Whereas Deluc was content to underline directionalism, Cuvier wanted to find changes both in the intensity and in the nature of former causes. He thus replaced transformation of species by catastrophic faunal changes.

In short, in Deluc's day, the late 1790s, the concept of uniformitarianism was considered to be an old method that had been used for decades and was based merely on common sense. How did Lyell turn it into a novelty?

## Innovative Uniformitarianism

Back from Sicily, Lyell told Murchison in 1829 that "*no causes whatever* have from the earliest time to which we can look back to the present ever acted but those *now* acting & that they never acted with different degrees of energy from that which they now exert."[20] He thus took the opposite view of Cuvier. While he pleaded his cause in the three volumes of his book, he also established a new methodology. The first volume of the *Principles of Geology* described current geological changes at the surface of the earth, from the effects of rain to those of earthquakes. Lyell emphasized processes "acting now" because they had been too much ignored by his contemporaries. Cuvier had rejected the doctrine of present causes in a few pages because he did not know much about it. Lyell, in turn, asked his readers to observe these processes, and he was certainly right to do so.

A simple confirmation of present-day causes was, however, not enough to create geology, as some biographers of Charles Lyell have too often stressed. A good illustration of the limits of the uniformitarian doctrine is the case of the zoologist and paleontologist Henri Ducrotay de Blainville (1777–1850) in France.

## Retrogressive Uniformitarianism

Blainville believed that "extinct groups have perished because of natural causes which are still active presently" and not by a "general revolution."[21] But because he naively believed in final causes, he could not admit that species changed according to circumstances, nor that they could have been created successively. According to him, the animal world formed an unbroken series (or a chain of beings) that did not allow piece-by-piece creation or repairs. For Blainville, only one solution remained: all species were formed at the same time, and fossil forms are those of species that have become extinct since Creation—a retrogressive thesis at a time when every paleontologist could observe that many present-day species were not found in the oldest rocks. Whereas Blainville's retrogressive uniformitarianism was only an unlucky offshoot of the doctrine, the weaknesses of Lyell's tectonics created even greater problems.

## Lyell on Mountains

Lyell did not seem to care much about the formation of mountains, or at least was unable to take a stand. It is surprising that, even in the tenth edition of *Principles of Geology*, he compared the formation of mountain chains to the general uplift of the Scandinavian countries.[22] This shocks the modern reader who knows that northern Europe is rising because of isostatic adjustment due to the melting of the ice cap that covered the region during part of the Quaternary. Based on the fact that the upper parts of the earth float on more or less viscous lower parts, the theory of isostasy says that the melting of ice has the same effect as the unloading of a ship. Such a movement cannot be compared to that which uplifts mountains by folding.

It is true that the idea of uplift in the northern countries impressed many naturalists at that time. Following Celsius and Linnaeus, the moving shorelines of the Gulf of Finland were attributed to the retreat of the sea. At the beginning of the nineteenth century, von Buch demonstrated the existence of uplifting movements.[23] But most naturalists remained skeptical. In 1834 Lyell took a trip to see for himself and was convinced.[24] Soon after, Élie de Beaumont presented a paper by Auguste Bravais (1811–1863) to the Academy of Sciences in Paris where he showed clearly that the ancient shorelines were deformed, a fact that could not be explained by the simple retreat of the sea.[25] Therefore, Sweden and Finland were in fact rising.

Of interest to uniformitarians was the fact that this movement was *slow*. If such a condition could be applied to any uplifting, then catastrophes would be excluded and mountain building could be explained by century-long movements. But how was it possible to explain this uplift?

Before the theory of isostasy was first presented in 1855 (see chapter 16), Lyell could at best explain slow uplifting by expansion of the earth's crust by heat. Heat and chemical reactions, he said, would "give rise to a mechanical force of expansion capable of uplifting the incumbent crust of the earth, and the same force may act laterally so as to compress, dislocate, and tilt strata on each side."[26] Catastrophists must have been baffled by such ignorance of the importance of folding. Murchison, Lyell's former travel companion, described in 1849 overturned rock masses and thrusts in the Glaris Alps that needed other types of lateral compressions.[27]

*Elements of Geology* proposed another solution, giving the impression that Lyell paid little interest to the question in the first place because he had a different answer ready whenever needed. However, this answer was not any more satisfactory. Having observed local collapse in coal mines, he concluded that "similar changes may have occurred at a larger scale in the Earth's crust."[28]

## Metamorphism

It was easy, however, for Lyell to triumph over earlier neptunistic ideas in regard to the origin of igneous rocks and the recognition of metamorphic rocks. Here rested probably the great novelty of the uniformitarian school.

Werner believed that granite and metamorphic rocks (gneisses, micaschists) were primitive rocks formed under conditions that no

## Metamorphism

*Metamorphism* means transformation of rocks at depth under the effect of temperature and pressure. Minerals of sedimentary rocks, formed at the surface of the earth, become unstable with increasing depth. During orogenic phenomena, increased pressure by lateral forces causes, in general, chemical reactions and structural changes of rocks. For instance, if the initial rocks are of argillaceous nature, they acquire a sheetlike structure that changes them into schists, micaschists, or gneisses. Phenomena of metamorphism are complex, and I shall abstain from greater precision.

The main metamorphic rocks (gneisses and micaschists) are both crystalline and schistose. This is why they are sometimes called crystalline schists. Their schistosity distinguishes them from granite. The former neptunists had classified them as primitive rocks, adjacent to granite. A close relationship between granitic massifs is in fact often found because the granitic magma originated (at least in part) from the melting of crystalline schists when temperature and pressure were higher than those required for metamorphism.

longer exist at the surface of the earth. Deluc had similar views, and his refutation of uniformitarianism was based on his observation that successive epochs contained deposits of different nature. Earlier, Buffon had subdivided the history of the earth into "epochs" according to the nature of the various rocks formed during each epoch. This neptunist viewpoint could thus be called in a global sense the "geology of epochs," or "periodical geology."

Hutton had opposed this view, saying that granite was an igneous rock that had risen by intrusion, uplifting and folding younger deposits. The plutonist school had thus presented a cyclic theory in opposition to a periodic one. Lyell accepted this theory and gave the name *metamorphic rocks* to sedimentary rock units that had been modified by the rise of magma.[29]

The idea of metamorphism had been envisioned by Hutton. Leopold von Buch accepted the idea of rising magma (of **pyroxene porphyry**) to explain the formation of **dolostones** by chemical transformation of limestones. Metamorphism was thus believed to be the result of the action not only of temperature but also of vapors.

At any rate, the notion of primitive rock was on its way out. In

1842, T. Virlet d'Aoust was able to say that "all rocks so far called primitive, may well be only of second, or third origin, if not of an even younger one."[30]

However, a very strict uniformitarian would not be entirely able to follow this author because Virlet, as well as Élie de Beaumont and the French school, distinguished two metamorphisms: one, called "normal," was caused by the action of a central fire upon the deepest rocks, that is, primitive rocks; the other, "abnormal," resulted from the heat of periodically injected intrusive igneous rocks. This concept thus included the two characteristics of catastrophism: directionalism of phenomena and periodicity. Normal metamorphism was directionalist, with variations in intensity of the phenomenon over time; whereas abnormal metamorphism was characteristic of earth's revolutions, which were periodically repeated.

Lyell, however, continued to believe in the uniformity of all geologic phenomena. Contrary to the cyclical geology of catastrophists, he envisioned an orogenic activity of constant intensity, but which moved from one place to the other on earth over time. His geology was indeed steady state according to the definition given earlier in this chapter. Lyell was neo-Aristotelian, and we can thus characterize his adversaries as successors of the Stoics—so much so that the directionalist doctrine introduced an evolutionist dimension. Indeed, the new geological cycle was evolutionary because it maintained Werner's idea of a succession of periods, thus inheriting the historical approach. Lyell, on the contrary, advanced on a uniformitarian base.

To separate clearly the fundamental views of catastrophists and uniformitarians, we should stress that three kinds of observations were involved. First, metamorphism transformed sedimentary structures as well as fossil remains. Metamorphism thus tended to erase the archives. Catastrophists believed that its action was moderate in modern times, whereas Lyell, and Hutton even more so, considered its role permanent.

Second, erosion was one of the mechanisms of the Huttonian cycle. Lyell followed this docrine, though he stressed marine processes whereas Hutton was particularly concerned with the erosion of lands. Erosion also erased part of the archives. In general, catastrophists minimized the effects of running water. Narrow valleys were often believed to be fissures resulting from uplift, collapse, or deformation of rock formations.

# The Didelphe of Stonesfield

Uniformitarianism called for an almost stable physical world and a biological universe with as little variation as possible to prove that animals could have lived in ancient times. However, since Cuvier it was known that fish preceded reptiles in the geological record and that reptiles lived before mammals. The question of the first appearance of mammals was therefore at the center of controversies about the uniformity of nature.

In 1812, fossil remains were found in calcareous shale at Stonesfield in England. Cuvier (1818) thought he recognized the jaw of a didelphe (etymologically meaning "two matrices," the term refers to marsupials who possess this anatomical characteristic). However, these rock layers were of Jurassic age, whereas it was generally believed that mammals had appeared at the beginning of the Tertiary era. The discovery thus supported the old age of mammals, namely, the uniformity of nature.

In 1824, Constant Prévost spent some time in England and sketched the jaw so that Cuvier could confirm his observation. However, Prévost was not sure that the fossil was of Jurassic age.

In 1831, Dufrénoy and Élie de Beaumont brought back a portion of a jaw, and everybody, including Cuvier, attributed it to a saurian. Thereafter, Ducrotay de Blainville reexamined the fossils and thought that they belonged to reptiles or fish, "which seems more in agreement with the age of the rock formation."*

It is known today that the jaws were those of authentic mammals—not of marsupials, but of forms, which lived before the separation of marsupials and real mammals (or eutherians), called panthotherians.

---

*Henri Ducrotay de Blainville, "Nouveaux doutes sur le prétendu Didelphe de Stonesfield," *Comptes Rendus, Académie des sciences de Paris* 7 (1838): 736.

The third observation concerned renewal of faunas. It was essential because renewal created archives, whereas the two other processes erased them. The theory of a gradual and uniform replacement of fauna supported by Lyell did not give much value to fossils. This is why it is fair to say that he was interested in stratigraphy in spite of his doctrine rather than because of it.

Drastic changes proposed by catastrophists had naturally quite opposite consequences. Catastrophists were therefore the real founders of biostratigraphy.

# Catastrophes That Built the World

## Directionalism

At the end of the eighteenth century, neptunism had established a history of the earth with the arrow pointing backward: lowering of sea level, depletion of the composition of the "liquid" (according to Deluc), weakening of active forces, and so forth. If these forecasts were correct, the earth was going to have no future. Catastrophists such as Élie de Beaumont proposed a directionalist model that was quite the opposite. They believed that ongoing geological processes were gradually building the earth. In other words, their model was uniformitarian, in a somewhat enlarged sense, insofar as it proposed that more or less permanent (or at least periodical) phenomena were building superposed structures, thus forming an increasingly complex organization.

At the basis of the idea was a process visibly lacking in neptunism, namely, orogeny (at the time simply called mountain building). Descartes and his followers pushed into the distant past the formation of mountains (by collapse). Even Moro, who talked about uplift, did not seem to believe that the process might repeat itself.

In 1829 Élie de Beaumont said, on the contrary, that nothing seems to "forecast a change of the Earth's crust which would assure us that the period of calm in which we live shall not be troubled by the appearance of a new mountain system, the effect of a new deformation of the land where we now live, whose present earthquakes forecast sufficiently that these foundations are not unshakeable."[1]

Deluc had also proposed repeated upheavals. But he respected the Bible too much to believe that such movements might repeat themselves on an earth created by God for the sake of humanity.

## Craters of Elevation

The first naturalist to decipher in the field the effects of successive uplifts seems to have been Leopold von Buch. Traveling through Italy and Auvergne at the beginning of the nineteenth century, he became convinced that basalt was of volcanic origin. Noticing volcanoes without craters (Puy de Dôme, Sarcouy), he was astonished by their shape and attributed it to a "swelling similar to a bladder."[2]

The idea of uplifting remained present in his mind when he traveled through Norway between 1806 and 1808. Although he admitted that the Scandinavian countries might have risen slowly, he still doubted that mountains could have been formed in such a fashion.

It was Grand Canary Island that served as a model for his famous theory of "craters of elevation." He believed that the rim surrounding the central peak was the broken center of a circular surface that had been uplifted. The peak itself was a volcano that had erupted through the opening.[3] In other words, von Buch believed that only the central peak was a volcano, whereas the uplifted area was of sedimentary origin. Thus, he attributed to an uplifting movement the circular surface, today called a volcanic **caldera**. One example is Vesuvius, which has two superposed volcanoes: one, culminating at Somma, erupted in A.D. 79 (destroying Pompeii and Herculaneum); the other, inside the caldera, was caused by later eruptions.

Nevertheless, volcanoes were not proper mountains for von Buch. He noticed that uplifting may form real mountain chains along faults. In 1823, in a letter to Humboldt, he wrote: "For some years now I have not doubted that the entire Alpine chain has been uplifted by the 'pyroxenic formation'" that fills there "some kind of huge vein whose direction is that of the mountain chain."[4] In the course of twenty years, von Buch had thus graduated from dome-shaped volcanoes to the origin of mountains. He cleared a last obstacle in 1824 when he distinguished in Germany four mountain systems with respect to their direction.[5] The search for ancient orogenies had thus begun. It became a major concern for Élie de Beaumont, who adopted similar ideas on some points while competing with others by von Buch.

## Élie de Beaumont

Élie de Beaumont was first known for his geological map of France (chapter 10). Professor of the School of Mines since 1829, he succeeded Cuvier in the chair of natural history at the Collège de France in 1832 and became permanent secretary of the Academy of Sciences in Paris after the death of Arago. An authoritarian scholar, Élie de Beaumont had an enormous influence on geology in the middle of the nineteenth century. Nobody actually dared to challenge his theories, but he had no real followers to continue his work. Those who criticize the dogmatic behavior of Werner may have forgotten that he at least had the privilege of seeing such brilliant students as Humboldt and von Buch contradicting him and establishing rival theories. Élie de Beaumont was much too rigid to have students of great talent.

In 1829, in the middle of fieldwork leading to the geological map of France, he read a long memoir at the Academy of Sciences, establishing "relationships of coincidence" between times of "uplifting of layers in some mountain systems" and gaps between "consecutive stages of deposit of sedimentary rocks."[6]

Before this, two different kinds of facts had been observed separately. On the one hand, Cuvier and Brongniart had shown the existence of "revolutions" characterized by changes in fauna (and flora). On the other hand, beginning with von Buch, mountains were being distinguished by their age. For Élie de Beaumont, it was tempting to relate these two facts and to think that revolutions were "the results of changes which had occurred along the seashores and in the sea itself by successive uplifts of mountains," mountain building having disturbed life in the sea.[7]

This was indeed an interesting idea. First, it provided a remarkable synthesis of two independent observations, thus giving coherence and unity to the emerging geology. Second, it explained biological changes that had remained mysterious events. More importantly, if one accepted that a more or less local uplift had worldwide repercussions on the sea, then the universal character of revolutions could be established. Long-range correlations of sedimentary series, which no longer rested on any theoretical basis since the abolishment of the neptunistic primitive ocean, could be justified. Replacement of faunas would have a general cause; they would therefore be synchronous all over the earth.

Furthermore, the theory could be tested. Uplift of a mountain chain could be dated by some "defect in parallelism," that is, an angular unconformity between older tilted layers and later horizontal ones. It was thus easy to note if this time corresponded to that of faunal renewal. To prove this was Élie de Beaumont's goal.

## Mountain Systems

In 1829 he recognized four "systems" (today known as orogenic phases). In 1833, when an abstract of his article appeared in the French translation of *A Geological Manual* (Manuel géologique) by Henry Thomas De la Beche, four systems had become twelve. In 1847 Élie de Beaumont found four new systems in the oldest rock units of Europe.[8] Five years later, he published *Notice sur les systèmes de montagnes* (On mountain systems). In this paper, an expanded version of an article he had been commissioned to write for a dictionary of natural history, he assumed there were twenty-two systems.[9] At the same time, he wrote to Constant Prévost that the number might be more than a hundred.[10]

It is evident that catastrophism would take on a moderate pace if the number of orogenic phases were multiplied. This is what Prévost also wanted to prove. If orogenic activity tended to become constant, then the distinction of separate phases lost its interest and precision. With much insight, Ami Boué, who had played a critical role all along because of the eclecticism of his nondogmatic thinking and his broad knowledge (he published both in French and in German), observed as early as 1831 that with an increasing number of uplifts, "we would find fewer errors in the use of the theory by M. de Beaumont because we would no longer have any means to vindicate ourselves by referring to intersecting uplifts."[11] A modern epistemologist might say that the system had become unrefutable.

## The Cooling of the Earth

What mechanism was behind these repeated uplifts? Élie de Beaumont proposed that they were the result of a "secular cooling" (in French: *refroidissement séculaire*) of the earth. In 1827 Louis Cordier (1777–1861), professor of geology at the Museum of Natural History, Paris, collected observations on the interior of the earth's crust at various depths. He demonstrated that temperature rose according to a certain gradient when one penetrates the earth's crust. Calculations made by Joseph Fourier (1768–1830) showed in 1824 that this gradient corresponded to a slow cooling of the earth.[12] Cordier believed that the crust, with a thickness of twenty-three to fifty-five leagues (one league corresponds to about four kilometers), was resting on a liquid matter.[13]

But Cordier drew only unimaginative conclusions about tectonics from his observations. Élie de Beaumont had a better idea. He stated that the earth's crust had reached its temperature of equilibrium and would not continue to cool off, but the interior of the planet continued to lose heat. As the interior cooled and contracted, the outer cover became too large for its shrinking core, collapsed, and formed folds.[14]

Earlier, Buffon had explored the concept of the earth's cooling. He built a model of the evolution of the earth that might be characterized as directionalist. But it lacked any good explanation of tectonics because Buffon associated mountain building with the change in state of the earth's crust from a liquid to a solid. Élie de Beaumont would link mountain building to a cooling of the planet, that is, to a continuous change in temperature; whereas Buffon's change of state was a unique event.

The theory of gradual cooling and contraction of the crust introduced into the process of mountain building a component of continuity that could satisfy the uniformitarians. Élie de Beaumont assumed, however, that collapse had occurred suddenly and that the movement had taken place during the interval of time that separated the most recent upturned layer from the overlying oldest horizontal one. This duration, he said, cannot be estimated. However, in other areas (where folding has not occurred), the same layers overlie each other continuously and are linked "by a more or less gradual transition. . . . The interval during which the observed stratigraphic unconformity took place was therefore extremely short."[15]

The uniformitarian Constant Prévost agreed with one point: Élie de Beaumont attributed mountain building to a movement of collapse of the earth's crust. Remaining faithful to Deluc's ideas, Prévost opposed any theory of uplift. Prévost noticed that Élie de Beaumont preferred to talk about folds rather than uplift. But Élie de Beaumont evaded the discussion altogether, even though he knew well that the old theory of collapse was obsolete.

He was all the more evasive because he understood the magnitude of shortening produced by collapse of the crust. In 1852 he noticed that some portions of the crust were folded and crushed as if caught between the jaws of a vise. He added that a block of rocks may overthrust (*chevaucher*) another, a term (used earlier by Dolomieu) that found its final expression only at the end of the century (chapter 15).

Another major point made by Élie de Beaumont was that the direction of mountain chains was extremely important—his principle of direction. To explain this odd theory today, in particular to students, the contraction of the interior of the globe is often compared to the drying of an apple, which becomes all withered, and the "wrinkles" of the earth's crust to that of the fruit's skin. This analogy suggests that mountain systems originated as haphazardly as the wrinkles of an apple. Élie de Beaumont's theory assumes just the opposite. He attached the greatest importance to the direction of chains and gradually constructed a very complex model known as *réseau pentagonal* (pentagonal network), discussed below.

## The Direction of Mountain Chains

In 1829 Élie de Beaumont specified that "the stratigraphic position of an angular unconformity is at the same level in chains which are parallel to each other," but it changes in chains "which do not have the same direction." Therefore, "deformations which have the same direction . . . were all produced by the same mechanism."[16]

This is an important point because, if the direction of chains and their age were linked, then the direction could be deduced from the age or vice versa; hence it was useless to define both. Because it is normally easier to find the direction of a chain, this could be done first. A geologist thus no longer needed to survey the whole mountain chain to find the location of an angular unconformity. The study of a geographical map could replace painstaking observations in the field.

According to Élie de Beaumont, the origin of his hypothesis could be found in Werner's observation that parallel faults were of the same age. Furthermore, Alexander von Humboldt (Werner's student), who was known for his narratives of travels in America and Asia and who founded botanical geography, had believed since 1792 that primitive layers at the surface of the whole earth present a uniformity of direction and of tilting. At first, he attributed these characteristics to a cosmological phenomenon, "a very universal cause based upon the first attractions which agitated matter."[17]

Later on, when Humboldt adopted the theory of uplifts, he began to have second thoughts. Finding tilted layers both in plains and in mountains, he wondered whether mountain chains had not emerged "similar to a chain of volcanic peaks" through "fissures formed parallel to the direction of preexisting tilted layers."[18]

It is obvious that Élie de Beaumont would have had to modify this concept to use it. In a way, Humboldt's idea played the same role as Hutton's concept of angular unconformities. As long as they were assumed to be universal, they could not be used to distinguish, by their age alone, uplifts from one region to another, or from a smaller range to another inside the same mountain chain. Élie de Beaumont had to render both discoveries *relative* before being able to include them in his theory.

Yet Humboldt's ideas on the direction of mountains and Hutton's concept of angular unconformities do not have the same status. The notion of angular unconformity and the principle of superposition were certainly principles of methodology. The notion that mountain chains of the same age run in the same direction was only a hypothesis subject to demonstration. Élie de Beaumont and his students deduced from the direction of mountain chains, established from maps, a hypothetical age that had to be verified in the field. Because the hypothesis was erroneous (as we know today), any observation in the field could only prove its fallacy. This would certainly have happened sooner had Élie de Beaumont not wielded his tyrannical powers over everybody around him, and had he not given ad hoc explanations when observations contradicted the dogma—such as, if the direction does not correspond to the age found in the field, then it must have "borrowed" the direction of a former uplift.

## The Pentagonal Network

The rigid geometrical character of Élie de Beaumont's theory of the pentagonal network finally undermined the theory itself. He postulated that uplifts of a same epoch occurred at the surface of a "spindle-shaped zone" following half of a great circle of the earth, similar to the "rib" of a melon. Each one of these zones represented, he said, a "system of mountains." However, the great half-circles were not scattered randomly; they were organized in a network of fifteen circles, crossing each other to create twelve pentagons on the surface of the earth. (A pentagon replaces, on a spherical surface, the hexagon, which, together with the equilateral triangle and the square, is the only regular geometrical figure able to divide a plane. On a sphere, equilateral triangles have angles of 60 degrees that cannot be assembled by six, as on a plane, to form hexagons.)

Mathematical constraints thus limited to fifteen the number of directions the great circles could take, and because the number of "mountain systems" was greater, it was necessary to assume "recurrent" directions, that is, that modern systems followed former systems.

Moreover, these mountain systems had to be dated, and Élie de Beaumont's efforts paralleled similar ones by stratigraphers who tried, at the same time, to divide rock units according to **hiatuses** (gaps) in the faunal succession, hiatuses that were generally interpreted as "revolutions."

Chapter 13

# The Help
# of Biostratigraphy

## The Division of Rock Units

Eighteenth-century naturalists first distinguished between primitive and secondary mountains and then defined transitional mountains as those located between the two. At the beginning of the nineteenth century, rock layers above the Chalk (which Werner had not distinguished from modern alluvials) were found to form a distinct unit, which was soon called Supracretaceous or Tertiary.

Thereafter, even smaller subdivisions were recognized according to their characteristics. For instance, in 1822 William Daniel Conybeare gave the name of **Carboniferous** to coal units in England. Chalk formations were named **Cretaceous** by J. J. Omalius d'Halloy in the same year. In 1929 Brongniart coined the word *Jurassic* to group together limestone units in the Jura Mountains. The term *Triassic* for rock formations older than the Jurassic was used as early as 1834. Finally, after 1835, Murchison and Sedgwick explored rocks below the coal formation.[1]

## The Stages

As stratigraphic data were increasing, geologists were trying to subdivide previously known systems. The term **stage** was coined to designate the new subdivisions. This notion was directly related to

| Rock Units | Stages |
|---|---|
| Contemporary | 28. Contemporary or present epoch |
| Tertiary | 27. Subapennine |
| | 26. Falunian {Upper Falunian / Lower Falunian or Tongrian} |
| | 25. Parisian |
| | 24. Suessonian |
| Cretaceous | 23. Danian |
| | 22. Senonian |
| | 21. Turonian |
| | 20. Cenomanian |
| | 19. Albian |
| | 18. Aptian |
| | 17. Neocomian |
| Jurassic | 16. Portlandian |
| | 15. Kimmeridgian |
| | 14. Corallian |
| | 13. Oxfordian |
| | 12. Callovian |
| | 11. Bathonian |
| | 10. Bajocian |
| | 9. Toarcian |
| | 8. Liasian (Liassic) |
| | 7. Sinemurian |
| Triassic | 6. Saliferian |
| | 5. Conchylian |
| Paleozoic | 4. Permian |
| | 3. Carboniferous |
| | 2. Devonian |
| | 1. Silurian {Upper Silurian or Murchisonian / Lower Silurian} |

Figure 13.1. Classification of Rock Units by A. D. d'Orbigny (From Cours élémentaire de paléontologie et de géologie stratigraphiques [Paris: Victor Masson, 1849–1852], 1:263).

the work of Alcide Dessalines d'Orbigny (1802–1857), who played an important role in the development of biostratigraphy.

With his division of rock units into Paleozoic, Triassic, Jurassic, Cretaceous, and Contemporary, d'Orbigny was able to subdivide rock units and recognize twenty-eight stages (fig. 13.1). His excellent knowledge of Jurassic fauna helped him divide that system into ten stages. However, in the Paleozoic (or Primary) rocks, he found only four stages—evidence of his insufficient knowledge of epochs that he had not studied personally.

It is, however, not d'Orbigny's lack of knowledge that is most disturbing for the modern reader, but his belief in the theory of multiple creations. In his *Cours élémentaire de paléontologie et de géologie stratigraphiques* (Beginning course of paleontology and stratigraphy), he stated that "each stage which has succeeded others through the ages of the world includes its particular fauna, set clearly apart from overlying or underlying faunas, and that these faunas did follow each other neither by transitional forms from species to species, nor by gradual replacement, but were separated by sudden destruction."[2] This is actually nothing new because Cuvier had been arguing for sudden destruction for a long time. We are not particularly surprised by d'Orbigny's opinion, in agreement with Élie de Beaumont, that "each time a mountain system was uplifted above the ocean, the existing fauna was destroyed by the extended movement of the waters."[3] But for his theory to be logical, d'Orbigny had to explain the appearance of new faunas.

## Multiple Creations

To account for new species appearing in the fossil record, Cuvier had merely appealed to migration. D'Orbigny assumed only a few catastrophes had occurred and differed in other respects from Cuvier's approach. However, the question may be asked today, If the twenty-eight successive faunas were actually *new* in the beds where they occurred, where had they been hiding before invading the sea one after the other? Therefore, d'Orbigny had to assume "successive creation of species in each geological epoch."[4]

It is true d'Orbigny added cautiously that "nothing can reveal to us the mystery involved" in these creations. Nevertheless, the mere idea of these creations bothered everybody at the time. Nonbelievers did not accept such repeated divine acts, and strict Christians were upset because they had difficulty believing in a God who had to make several attempts at Creation. The difficulty increased if one assumed, as Marcel de Serres (1783–1862) suggested, that past creations have grown more and more perfect and that they will continue to do so in the future. Consequently, humans could not be "the goal and masterpiece" of divine Creation.[5]

Contrary to many geologists of his time, d'Orbigny did not believe in progression. He energetically rejected the law of "gradual improvement of living beings, from ancient times to the present."

Nevertheless, his name, more than anyone else's, has remained asso-ciated with the theory of successive creations. His colleagues in the Geological Society of France were so embarrassed by his ideas that his eulogy was only written twenty years after his death in 1857, once "the raging debates stirred up by his doctrines had become extinct."[6]

## The Transformist Explanation

Two years after the death of d'Orbigny, the problem of faunal succes-sion was solved by the Darwinian theory of evolution. This solution raised embarrassing questions, however, because it not only stirred up religious objections, but it also assumed a very strict continuity (gradualism) of faunas, which no paleontologist had yet found in the layers of the earth. Nevertheless, as soon as transformism began to be accepted (whether for biological or ideological reasons), continuity was stressed, and biostratigraphy, with increasingly precise data, produced examples of species that had passed from one stage to the next or were found to be unchanged over great spans of time—species of "living fossils," which had existed tens or hundreds of millions of years.

## Facies

Groups or organisms that evolved slowly (or not at all) became im-portant for the distinction of paleontological facies. Beginning in 1836, the term was used by Amanz Gressly to describe two major situations: "The first one assumes that a given petrographic aspect [lithofacies] of any rock unit, wherever it is found, is necessarily as-sociated with the same paleontological assemblage [**biofacies**]; the second assumes that a given paleontological assemblage excludes genera and species of fossils which abound in other facies."[7]

In other words, it was recognized that some living organisms cor-respond to a particular environment (an ecologist would say that this environment, or biotope, is inhabited by a particular assemblage or **biocenosis**). As a result, it was possible to distinguish two kinds of fossils useful for geologists. **Index fossils** are genera, rarely species, that evolved rapidly and were distributed widely; they are used to date rock layers. **Facies fossils,** usually a single species or genus, lived within a restricted area and are used to reconstruct past geographies. Facies fossils are particularly important when they still have modern representatives, because it is then possible to know their environment. Therefore, organisms with very slow rates of evolutionary change are also of great interest.

It seemed obvious that, to be so similar, ancient and modern forms must have lived in the same environment—an assumption that could lead to error. For example, when fossil proboscideans (an order of mammals that includes the elephant) were discovered, Buffon believed that all proboscideans had lived in tropical countries and inferred that present temperate regions once had torrid climates. It became known later on that mammoths had lived in cold regions. Zoological closeness between species does not necessarily imply a similarity in their mode of life. Even anatomically stable species may have changed their environment in response to competition from surrounding species. Living fossils, which have not changed through the ages, often live in very restricted environments compared to their initial habitat. They are confined species.

The idea of facies fossils goes back to Woodward, who in 1695 distinguished **pelagic** from **littoral** forms. He forecast the course taken later by Rouelle and Lavoisier. Lavoisier had plans for a more systematic study of the relation between water depth and marine organisms when the French Revolution ended his life.

The study of fossils as indicators of environments was more easily acceptable for some naturalists than their use for dating rocks because it did not require a hypothesis on the life and death of species; such studies could be done first with "easy" fossils, those resembling present species. The only obstacle at that time was poor knowledge of pelagic faunas. This is why marine explorations in the nineteenth century contributed so much to the development of the study of facies.

## Cumulative Knowledge

With these new tools, stratigraphy was able to develop. Knowledge had become cumulative. D'Orbigny played the same role in developing the outline of a geological time scale that Élie de Beaumont had in world tectonics.

If we remember only erroneous interpretations in their respective works—the pentagonal network or multiple creations—we clearly commit an injustice, or worse, an error of perspective. Indeed, these theories, though proven wrong, have provided the basis for two great fields in modern geology: stratigraphy and tectonics.

In the introduction to this book, I compared the great number of eighteenth-century theories of the earth to the unifying ideas of nineteenth-century geology. After Werner and Hutton, concepts and methods that allowed accumulation of knowledge were established. Although the proposed theories were still subject to revision—the basis of scientific methodology—their fall would not require a drastic reexamination of previously acquired knowledge.

This is shown by the increase of manuals and treatises in the middle of the nineteenth century. Toward 1810, dictionaries of natural history were published, presenting everything that was known in alphabetical order. Some ten years later, didactic works gathered knowledge and arranged it by subject matter, as, for example, Charles Lyell's works or *A Geological Manual* by H. T. De la Beche (1832). D'Orbigny's *Cours élémentaire* was the prototype for stratigraphy.

## Contemporary Stratigraphy

It is impossible to give all the details about the various stages in the making of stratigraphy after d'Orbigny's death. Most works attempted not only to improve the system but also to unify it because stratigraphic scales were different in each country. To correlate rock units for the entire earth, a unique chronometer was necessary.

In 1874, Eugène Renevier, professor at the University of Lausanne, synthesized the various attempts at dividing rock units into stages in his "Tableau des terrains sédimentaires" (Table of sedimentary rock units).[8] He published a corrected edition twenty years later with the eloquent title "Chronographe géologique" (Geological chronograph).[9] In the meantime, the first International Geological Congress was held at Paris in 1878, where the various viewpoints were discussed. Today, about eighty stages are distinguished.

# Unraveling the Earth's Age

## Division of Time

In 1829 Alexandre Brongniart published *Tableau des terrains qui composent l'écorce du globe* (Table of rock units which form the earth's crust), in which he proposed dividing geological time into two periods: a Saturnian period during which continents were more or less covered by the sea; and a Jovian period corresponding to the time when continents began to acquire their present shape.[1] The first included all the marine transgressions and **regressions** since the beginning of time up to the last one, and the second period referred only to that most recent regression.

Brongniart's classification, inherited from J.-A. Deluc and Cuvier, was very egocentric (or anthropomorphic) because it treated modern times in a very subjective fashion. However, every history favors the present over the past, if only because of the unequal amount of data available for the two.

A historian of the earth, no less than the chronicler of human history, cannot escape curtailing the past, simply because so little is known about older rocks. For instance, modern terminology uses *era* to designate spans of time that are increasingly long the more remote they are from the present. For instance, the Cenozoic era is 65 million years long, the Paleozoic era, 345 million years, and the Precambrian era at least 4,000 million years. Furthermore, common speech (in comic strips, for example) often lumps all geological eras into one "prehistoric" or even antediluvial period. An objective division of the history of the earth can only be achieved in a distant future, if ever, when our knowledge of the strata of all ages will be known with the same degree of accuracy.

During the 1830s and 1840s, the exploration of older epochs provided an excellent incentive to get to know rocks that had never been studied before. Geologists were busy describing rocks older than the coal-bearing strata.

## British Geologists

It is not surprising that the oldest rocks of the Paleozoic were first studied in Great Britain. Indeed, the old massifs in France that represent the basement (chapter 6) date from the Hercynian orogeny (second half of the Paleozoic), whereas the greatest part of the British Isles is older, dating back to the **Caledonian orogeny,** at the beginning of the Paleozoic. Hence, induration of rocks in Great Britain preceded the same process in France. As a result, the Devonian (whose age is between these two orogenies) forms in England a horizontal deposit represented particularly by the Old Red Sandstone, a very thick sequence (several thousands of meters) of lacustrine (formed at the bottom or along the shore of a lake) or fluvial rocks. Its equivalent in France was highly folded during the Hercynian orogeny.

At the beginning of the nineteenth century, several British geologists played an important part in the unraveling of the Devonian. The ensuing controversy has been studied in detail by Martin J. S. Rudwick.[2]

Adam Sedgwick (1785–1873) was a professor of geology at the University of Cambridge starting in 1818. His friend and co-worker, Roderick Impey Murchison (1792–1871), remained an amateur and country gentleman for some time. Encouraged by his wife, he finally decided to become a scientist. He chose geology because it allowed him to continue outdoor exercise and hunting. He attended lectures in geology given by William Buckland (1784–1856) at Oxford.

Buckland was first appointed reader at Oxford because there was no chair of geology at that famous university, which apparently did not favor such scientific studies. In 1818 he petitioned, with success, for the additional title of Reader in Geology. Every geologist of the generation of Murchison and Lyell was a student of Buckland, whose merits as a teacher were highly praised. His inaugural lecture rehabilitated the close relation between geology and Genesis.[3] Rudwick wrote that this was "partly an attempt to reassure Oxford colleagues that the pursuit of geological science would not subvert Christian faith and piety, and that it was therefore a proper subject to be taught at the university that was the intellectual center of established religion in England."[4]

## Cambrian and Silurian

Toward 1830, little was known about rocks older than the coal-bearing strata. Abundant **detrital rock** units found underneath the coal were simply named the Old Red Sandstone, but strata lower yet remained a confusing mass of so-called transition rocks. Henry Thomas De la Beche (1796–1855), the first director of the Geological Survey of Great Britain and author of teaching manuals (chapter 13), divided older rocks into **grauwacke** and lower fossiliferous group. Murchison, in turn, began the study of these rocks in western England and Wales. He described a series of four formations characterized by their faunas, which he interpreted as underlying the Old Red Sandstone. In 1835 he named these formations **Silurian** in honor of a Celtic tribe who had fought valiantly against the Romans. In 1839 he published a synthesis entitled "The Silurian System."[5]

Sedgwick arrived at a different conclusion. He was more influenced by tectonic events than by faunas and described rocks with few fossils, which appeared to him older than those mentioned by Murchison. He proposed the name **Cambrian** (Cambria is the Latin name for Wales) for rocks that included Murchison's Lower Silurian.[6]

Élie de Beaumont provided a synthesis by restricting the term Cambrian to rock units older than the Silurian, as defined by Murchison. To avoid all ambiguity, he replaced the name Cambrian with Cumbrian, which remained in use for some time before it was changed back again to Sedgwick's term.

Simultaneous studies showed that the fauna of rocks in Devonshire was in part very similar to that of the Silurian and in part to that of the Carboniferous, whereas a third portion appeared to be unique. After further studies by Murchison and Sedgwick, it was concluded that these formations were equivalent to the Old Red Sandstone in Scotland, southern England, and Ireland. Because this marine fauna was richer than that of the continental Old Red Sandstone, it was given the name Devonian, a system between the Silurian and the Carboniferous.[7]

Shortly after, the Permian, the last period of the Paleozoic era (with deposits of the New Red Sandstone), was named by Murchison in 1841 when corresponding rocks were found in the Russian government or territorial area of Perm. With the exception of the **Ordovician,** all the systems of the Paleozoic era were thus named and described. Therefore, "transition rocks" of the Wernerian school no longer represented a mysterious ancient world that was subject to conditions unrelated to those of the modern world.

It is true that rocks even older than the Cambrian (or the Silurian for authors who refuted Sedgwick's division) existed. They were called **Precambrian.** However, the term did not catch on in the 1840s. Murchison and French geologist Édouard de Verneuil (1805–1873) preferred the term **Azoic,** referring to an environment devoid of life. Indeed, the great problem now was determining when life on earth began.

## Primitive Fauna

The solution to that problem would allow the writing of the history of the earth in a direct time order. Werner, who had been sure— although today known to be wrong—that the earth's first state was a primitive ocean, had told its history in a direct time order too. He started at the bottom of strata and went up through successively younger ones. However, his followers cautiously retreated from this ambitious goal. It became a habit to start with the present and proceed in reverse time order—"from the known to the unknown," as Ami Boué used to say. Indeed, in 1830 naturalists preferred the inductive method because inductive reasoning and uniformitarianism (with its central concept: the present is the key to the past) were in fact two facets of the same approach.

Murchison returned to the direct time order in 1839 when he believed he had a new starting point. According to B. Balan, the time was ripe to accept "a positive time direction which is first of all the direction of gradual development of life, starting from a base level considered as initial."[8]

Shortly afterward, Joachim Barrande (1799–1883) continued his search for what he called the "primordial" fauna. Educated in Paris at the Polytechnic School and the School of Civil Engineering, he became tutor to the grandson of Charles X and left France with the royal family in 1830. After some travels, he remained in Prague and became interested in local fossils such as **trilobites.**

In his huge work *Système silurien du centre de la Bohème* (The Silurian system in the central part of Bohemia), he stated that the rare fossils found in rocks older than the "primordial fauna" are sporadic forerunners of that fauna; that there is not much hope of finding fossils numerous enough to earn the name of fauna in rocks that we call azoic; and that the present fauna is the richest.[9] According to Barrande, starting from the present level and tracing these fossils backwards through the geological record, we will see some of the most highly organized types disappear, bed by bed, so that by the time we reach the level of the primordial fauna we find only one well-developed family, that of trilobites, and some lonely species as sporadic forerunners of mollusks and radiata.

Before Barrande, it was believed that the absence of organic remains in the oldest rocks could not necessarily prove that life did not exist as yet simply because metamorphism had made these rocks (called metamorphic schists after the dismissal of the adjective "primitive") unreadable. Of course, the numerous supporters of "normal metamorphism," who claimed that these rocks had been metamorphosed during their deposition, could add that life was incompatible with the conditions of such a transformation. However, none of these arguments could match that by Barrande who, upon discovery of the primordial fauna, established an origin of life. His thesis became all the more plausible when geologists started to explore weakly metamorphic Precambrian rocks and discovered that they were indeed azoic.

However, as always when things become very probable, when facts prove the most solid theories, exceptions appear, accumulate, and finally invalidate the system. In 1858, remains of what appeared to be organisms were found in Precambrian rocks on the banks of the Ottawa River in Canada. John William Dawson (1820–1899) at McGill University in Montreal carefully examined these remains and concluded that "these remarkable structures are truly organic." He classified them among foraminifers (microscopic unicellular animals with calcareous tests—hard protective shell or cover of certain invertebrates), named them *Eozoön canadense* (1865), and proposed replacing the name Azoic with Eozoic to designate Precambrian times.[10] The prefix eo- stems from the Greek ēōs, dawn or morning, hence, the dawn of life.

But Dawson was wrong. Formed by tiny superposed layers of calcite and serpentine, connected by pillars, these purely mineral structures were indeed found later in limestone blocks on Monte Somma, ejected by eruptions of Vesuvius after having been torn off from Jurassic layers of the substratum of the volcano.[11] Nevertheless, the search was on, and in the following years an authentic Precambrian fauna was discovered.

In 1894 Charles Barrois collected microorganisms in old rocks at Lamballe, Britanny.[12] Thereafter, such organisms were found in graphitic layers between metamorphic schists in Finland. Today, numerous types of Precambrian fossils are known. Besides the two most abundant types, bacteria and algae, the latter forming reefs called **stromatolites,** relatively complex organisms have been found: jellyfish, worms, and others.

## Absolute Time of the Catastrophists

With increasing knowledge about earlier times and further exploration of the past, questions about the absolute geological age of these rocks, as well as that of the earth, became more and more acute. For a long time, only indirect and unreliable methods existed to answer these questions.

Stratigraphers who studied successively superposed rocks had determined their relative age so that they could be placed in a column, as geognosists had done earlier. However, they were still unable to give their duration or their age in millions of years. Naturalists had long tried to make up for that deficiency by various means. For instance, Buffon estimated the duration of the earth's cooling by comparison with that of cannonballs. He also calculated the rate of deposition of shales based on the opinion that a yearly deposit of such a sediment could be no more than five feet. Others calculated the rate of erosion. Gautier, for instance, measured "how much mud was transported by rivers from land to sea each year, and he deduced the time necessary to erode continents."[13] Palassou stated that one million years were needed to destroy the Pyrenees.[14] Other naturalists—from Maillet to Linnaeus—supported the theory of the diminution of the sea, making calculations that were just as speculative.

In the nineteenth century, time was at stake in debates about the action of supposedly slow present-day causes. Recall Lyell, for example, who was led to such theories after his guess at a geological time scale. It would be, however, too simple to conclude that every catastrophist was in favor of short chronologies, even if they did not still subscribe to the seventeenth-century belief in a global history of a scant six or eight thousand years.

Cuvier, for instance, did not hesitate to talk about "thousands of centuries" when referring to the past of the earth in his *Discours* (1812). However, in subsequent editions he omitted any reference to time. In 1838 Marcel de Serres talked about millions of years.[15]

The British were even more daring. Buckland, who wanted to demonstrate "the wisdom and the benevolence of God shown in his works of Creation," talked about "millions and millions of years."[16] His friend William Conybeare set a record by estimating the age of the earth at "quadrillions of years."[17]

We should not be astonished at the audacity of some catastrophists. The durations of deposition measured geological time—and almost everyone agreed on that point—but the durations of orogenies and of biological "revolutions" did not. Catastrophists placed orogenies between phases of deposition, whereas uniformitarians placed them during these phases. Hence, catastrophists did not have to assume shorter durations than their adversaries.

Nevertheless, all chronologies of directionalists had one major flaw. Paradoxically, they were not too short, but too long. We all know that anything exaggerated is not reliable. To come up with numbers as enormous as they were unjustified meant that naturalists were not truly concerned with measuring geological time.

## Time According to Uniformitarians

Lyell, on the contrary, tried to estimate time without indulging in too-risky speculations. He compared faunal changes in the Quaternary with changes during earlier periods. He noticed that 5 percent of the fauna had changed during the glacial period. Assuming a constant rate of renewal (a uniformitarian hypothesis), he calculated that twenty times more time was necessary for a biological "revolution"

---

## Sedgwick, Buckland, and Murchison

In his autobiography, Charles Darwin referred to Sedgwick, Buckland, and Murchison in the following terms:

Professor Sedgwick intended to visit N. Wales in the beginning of August [1831] to pursue his famous geological investigation amongst the older rocks, and Henslow asked him to allow me to accompany him. Accordingly he came and slept at my Father's house.

A short conversation with him during this evening produced a strong impression on my mind. Whilst examining an old gravel-pit near Shrewsbury a labourer told me that he had found in it a large worn tropical Volute shell. . . . I told Sedgwick of the fact, and he at once said (no doubt truly) that it must have been thrown away by someone into the pit; but then added, if really embedded there it would be the greatest misfortune to geology, as it would overthrow all that we know about the superficial deposits of the midland counties. These gravel-beds belonged in fact to the glacial period, and in after years I found in them broken arctic shells. . . .

All the leading geologists were more or less known by me, at the time when geology was advancing with triumphant steps. I liked most of them, with the exception of Buckland, who though very good-humoured and good-natured seemed to me a vulgar and almost coarse man. He was incited more by a craving for notoriety, which sometimes made him act like a buffoon, than

---

to occur. Lyell counted four revolutions since the end of the Cretaceous and eight others for earlier times. Since his contemporary James Croll (1821–1890) had estimated—based on astronomical calculations—that the glacial period had lasted one million years, Lyell concluded: "If each, therefore, of the twelve periods represents twenty million years on principles above explained, we should have a total of two hundred and forty millions for the entire series of years which have elapsed since the beginning of the Cambrian period."[18] (Radioactive dating of rocks in the twentieth century doubled this number so that the Cambrian began in reality 570 million years ago.)

At the end of the nineteenth century, sudden doubts arose about the old age of the globe. The distinguished physicist Lord Kelvin (William Thomson, 1824–1907), known for his work in thermodynamics (Kelvin scale), calculated, on the basis of present heat flow at

*(continued)*

by a love of science. He was not, however, selfish in his desire for notoriety; for Lyell, when a very young man, consulted him about communicating a poor paper to the Geol. Soc. which had been sent him by a stranger, and Buckland answered—"You had better do so, for it will be headed, 'Communicated by Charles Lyell,' and thus your name will be brought before the public."

The services rendered to geology by Murchison by his classification of the older formations cannot be overestimated; but he was very far from possessing a philosophical mind. He was very kind-hearted and would exert himself to the utmost to oblige anyone. The degree to which he valued rank was ludicrous, and he displayed this feeling and his vanity with the simplicity of a child. He related with the utmost glee to a large circle, including many mere acquaintances, in the rooms of the Geolog. Soc. how the Czar Nicholas, when in London, had patted him on the shoulder and had said, alluding to his geological work—"Mon ami, Russia is grateful to you," and then Murchison added rubbing his hands together, "The best of it was that Prince Albert heard it all."*

---

*Charles Darwin, *The Autobiography of Charles Darwin: 1809–1882*, with original omissions restored, edited with appendix and notes by his granddaughter Nora Barlow (New York: Harcourt, Brace and Co., 1958), 69, 102–103.

the surface, that the earth could not be more than 100 million years old and was likely much less.[19] This time limit reduced the duration of biological evolution. This is why Thomas Huxley (1825–1895), a friend of Darwin's, disputed Kelvin's calculations.

Nevertheless, Kelvin's results did impress geologists who had become accustomed to counting on longer durations. The controversy ended when John Joly (1857–1933) published *Radioactivity in Geology* in 1909 and showed that the earth's heat resulted not only from its initial heat, but also from heat produced by radioactivity.[20] The accidental discovery of natural radioactivity by Henri Bécquerel in 1896, followed by the important studies done by Pierre and Marie Curie on the origin of radioactivity, reversed the results obtained by Kelvin.

It was found that helium, which was known to exist in the sun, occurred on earth as a product of radioactive decay of uranium. Lord Ernest Rutherford (1871–1937) studied the relationship between helium and uranium in minerals. Starting in 1917, it became possible to know absolute geological time.

Of the many radioactive isotopes that exist in nature, rubidium 87, strontium 87, and uranium 235 and 238 are used to date rocks that are millions of years old, whereas potassium 40 and argon 40 date rocks as young as fifty thousand years. Carbon 14 is used to date events of even more recent geological history.

Radioactive isotopes are not only useful for measuring geologic time: "Recent advances in nuclear physics have allowed new research in geochemistry; its technical and practical importance results from the distribution of isotopes which can explain dispersion and concentration of chemical elements."[21]

# The Breaking Up
# of the Crust

## Eduard Suess

Murchison, Sedgwick, Élie de Beaumont, and Lyell all died between
1871 and 1874. Prévost, De la Beche, Buckland, and d'Orbigny had
left the stage twenty years before. At the end of the 1870s the founders of geology were all gone, and a new generation took their place.

In 1875 a work on the origin of the Alps, *Die Entstehung der Alpen* (The formation of the Alps) by Eduard Suess (1831–1914) was
published.[1] Born in London, he lived thereafter mostly in Vienna,
where he began his scientific career as a paleontologist. In 1857 he
was appointed professor of paleontology at the University of Vienna,
and later on professor of geology. Soon he was the undisputed master
in that field. Suess was active in many other fields, including poli-
tics. He became a member of the municipal council of Vienna and
was later deputy at the assembly, where his left-wing oratorial skill
made him a formidable adversary.

In 1878 Suess began the preparation of a gigantic work, *Das
Antlitz der Erde* (The face of the earth), the first volume of which appeared in 1883.[2] The undertaking was of such magnitude that it took
twenty-six years of the author's life; part two of the third and final
volume was published only in 1909.

Suess was at first interested in present-day movements of the external crust of the earth, a sign that the uniformitarian method was

by then well integrated into science, even for a geologist who, as we shall see, accepted violent and sudden events.

Strangely enough, the first process of present-day natural phenomena to attract Suess's attention was the Deluge. The author recalled that, besides the biblical account of Noah's Flood, Mesopotamian texts related a similar event. Suess quoted, in particular, Assyrian writings and the account of Berosius, a Babylonian priest who lived in 300 B.C. Suess interpreted these accounts by hypothesizing a powerful earthquake associated with a hurricane that pushed the waters of the sea upstream through the Euphrates Valley. He explained that the biblical version differed from his interpretation because it was written by people, located far away from the event, who distorted the narrative. Suess concluded that the Deluge was not a global event and that Mount Ararat was not the summit generally thus called today, but a small hill with the same name.

Suess wanted to show that, besides the more regular movements of the earth's crust—the only ones considered by uniformitarians—exceptional and powerful processes could contribute to shape the "face of the earth." He decided to enlarge the field outlined by Lyell and to undertake a synthesis of uniformitarianism and catastrophism.

## The Earth under Tension

Suess adopted a clearly directionalist attitude. Like Élie de Beaumont he believed in the cooling of the earth, to which he attributed deformation of the crust. Contraction generated by cooling created a tension between crust and interior of the globe.

He wrote that two types of tensions exist: first, tangential ones (parallel to the surface), generating lateral thrusts that fold beds and form mountains; second, radial ones (perpendicular to the surface), which produce collapses. Like Élie de Beaumont, Suess believed that the essential movement was the shortening and collapse of the crust.

Collapse structures may be polygonal in shape, bounded by cracks called peripheral, and cut across by radial cracks (radiating from the center). The quadrilateral basin of Bohemia, he said, fit this description perfectly. The Red Sea, the Jordan Valley, and Alsatia are collapse structures of the same type. Circular collapse structures, devoid of radial cracks, are found in the Bay of Naples.

Tangential tensions, which produce mountain chains, result from thrusts that push beds to override a **foreland**. The inevitable consequence is an asymmetry of the chain. In 1875 Suess stated that the Alpine system originated from a thrust from the south or southwest, thereby stressing the importance of tangential movements.

However, radial and tangential movements were not independent, he said. Often the lateral thrust that produced the chain was associated with collapse of the region located behind it, which Suess called **hinterland.** In the Alps, for instance, collapse produced the Adriatic Sea and the western Mediterranean Sea. The foreland may also collapse; for instance, Bohemia and Alsatia belong to the Alpine foreland.

So long as mountain building was still believed to result from the contraction of the globe, Suess's synthesis, despite its new descriptive criteria, was comparable to that of Élie de Beaumont, Suess's predecessor. However, Suess abandoned one major concept. He observed that mountain chains display curved trends and that it is therefore not possible to relate the age of folding to the direction of chains according to Élie de Beaumont's principle of direction. Suess set himself even more apart from his predecessor when he proposed that Europe had been formed by a series of successive orogenies, thus replacing the mountain systems of the previous school of thought. Suess motivated a younger French geologist, Marcel Bertrand (1847–1907), who immediately realized the novelty and potential of these revolutionary ideas.

## Marcel Bertrand

Marcel Bertrand studied at the French Polytechnic School and the School of Mines and graduated as an engineer in 1872. He was appointed to a position at the survey, which was preparing a detailed geological map of France (begun in 1868 at the request of Élie de Beaumont). Until then, Marcel Bertrand had shown but little interest in geology. No wonder; Élie de Beaumont, his teacher, is said to have been a rather mediocre one, talking in a low voice with his back turned toward the class.

Suess's *Die Entstehung der Alpen* fired Bertrand's enthusiasm. He understood that the key to the great enigmas in geology could be found in the Alps. He studied first the Jura Mountains and then the Provence. In 1884 he astonished the Geological Society of France with his paper on the structural relationships between the Alps of Glaris and the French-Belgian coal basin. He demonstrated that certain abnormal contacts between rock units of different ages could be explained by large-scale overthrusts, today known as **nappes.**[3]

This meant that older rock units could be displaced by a powerful horizontal thrust so as to override younger ones. The result is an abnormal order of superposition with younger rocks located below older ones, thus contradicting the old principle of Steno. In the French-Belgian coal basin, for instance, Carboniferous rocks are overthrust by older rocks of the Devonian or Silurian period. The Alps of Glaris in Switzerland display a similar example of Permian rocks thrust over Tertiary ones.

Overthrusting of older rocks on younger ones naturally assumes lateral thrusts, a concept in agreement with Suess's horizontal displacement theory. Consequently, Bertrand read with great interest the first volume of *Das Antlitz der Erde,* published in 1883.

When Marcel Bertrand became professor at the School of Mines, he gave an important lecture on March 21, 1887, entitled "La chaîne des Alpes et la formation du continent européen" (The chain of the Alps and the formation of the European continent).[4] In this lecture he developed Suess's ideas and added the consequences he drew from them. The most important pertained to continuity of folding.

## The Principle of Continuity

Bertrand distinguished "at least four indisputable major movements": in the Cretaceous, in the Eocene, and in the Middle and Late Miocene. These "movements built one single chain," the Alps. "Therefore, the question may be asked if these movements had been really well defined, as postulated by the old theory, occurring at certain predictable times, and affording precise dating for the distinction of geological periods, or if, on the contrary, these movements had prevailed in a continuous and uninterrupted manner during the entire length of Mesozoic and Cenozoic times."[5]

Bertrand thus criticized Élie de Beaumont's synthesis, whose unifying factors were the identity of direction of chains of same age and the division of the history of the earth into epochs, separated by violent catastrophes. Bertrand showed the uniformitarian character of orogenies, but preserved the concept of cyclicity in the process. He replaced Élie de Beaumont's principle of direction with a principle of continuity in which the unifying factor was continuity of formation of each mountain range.

The assumed mountain systems became henceforth mere epi-sodes in the construction of mountain ranges, and the formation of Europe was more likely a succession of several juxtaposed chains rather than a discontinuous sequence of systems. Consequently, Marcel Bertrand started to identify the various chains forming the European continent.

He wrote that the latest chain corresponds to the Alps. It can be followed "from Switzerland to Austria and the Carpathians." The Alps were preceded by a chain he called Hercynian—using a term coined earlier by Leopold von Buch—"which cuts across Europe obliquely from Silesia to southern Ireland." The movements that formed this chain took place at the end of the Paleozoic era. Today, rocks folded at that time are largely buried under a Mesozoic and Cenozoic cover (see chapter 6). They crop out only in the massifs of Central Europe (Armorican Massif, Central Massif, Vosges), Bohemia (Harz, from which the name of the chain originated), Black Forest, and Cornwall.[6]

The French-Belgian and Ruhr coal basins belong to this chain. This is why Bertrand at first called it "chaîne houillère" (coal-bearing chain), whereas Suess named it the Variscan chain after an ancient tribe of Saxony and Bavaria.

To the north of the Hercynian chain, Devonian and Carboniferous rocks remained horizontal (the continent of the Old Red Sandstone, see previous chapter). Their pre-Carboniferous substratum must have been folded earlier. Because it extends over Norway and Scotland (ancient Caledonia), Suess and Bertrand named this structural unit formed at the beginning of the Paleozoic the Caledonian chain.

Marcel Bertrand expanded this structural model, comparing the Allegheny Mountains in North America to the Hercynian chain, and the Green Mountains to the Caledonian chain. These chains there-fore crossed the Atlantic. Because Canada shows structures older than the Paleozoic, Bertrand proposed the existence of an Archean continent to the north of these zones. Consequently, the European and North American continents were formed by three successive orogenies extending along the southern margin of an old continent, which henceforth gradually increased in size (fig. 15.1).

Élie de Beaumont's pentagonal network of mountain systems was thus replaced by Bertrand's theory of continents accreting around older nuclei by juxtaposition of marginal bands. Bertrand believed that continents are not permanent, pointing as evidence to the

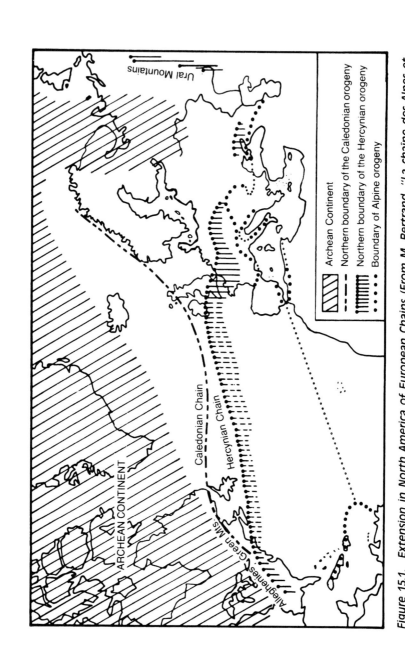

Figure 15.1. *Extension in North America of European Chains (From M. Bertrand, "La chaîne des Alpes et la formation du continent européen," Bulletin Société géologique de France, 3d ser., 15 [1887]: 442).*

Legend:
- Archean Continent
- Northern boundary of the Caledonian orogeny
- Northern boundary of the Hercynian orogeny
- Boundary of Alpine orogeny

Map labels: Ural Mountains, ARCHEAN CONTINENT, Caledonian Chain, Hercynian Chain, Alleghenies (Green Mts)

present Atlantic Ocean, which occupies the center of the European–
North American continent. Moreover, Suess revived the old collapse
theory because portions of continents collapsed, he said, generating
troughs of variable size. Thus, his contemporaries might have argued
that Plato's Atlantis was not a myth, because the Austrian geologist
believed that the Atlantic was of Miocene age, albeit somewhat older
than the appearance of humans.

The concept of sudden collapses reestablished long-distance correlations that had been threatened with disappearance because of
some geologists' ideas on continuous movements of the crust. I mentioned earlier that global catastrophes were of great interest to those
who wished to explain simultaneous faunal changes from one continent to another. If tectonic processes were continuous they could
have no effect on evolution, which would therefore become more or
less independent. In other words, only sudden catastrophic events
would have global evolutionary repercussions. Stratigraphers were
probably not catastrophists by trade, but if they started to believe just
a bit in general revolutions, they could begin to question the correlation between their datings from one region to another.

## Repeated Collapses

In 1888 Eduard Suess published the second volume of *Das Antlitz
der Erde,* in which he discussed processes of marine transgressions
and regressions. Following the uniformitarian approach, he began to
study the famous problem of moving shorelines, which from Celsius
to Lyell had preoccupied generations of scientists.

Leopold von Buch and his successors had decided in favor of uplifting of the continent. Suess challenged these ideas. He believed
the problem to be related to climatology rather than to tectonics
because the Baltic Sea opens into the North Sea. For him, the movement was a negative movement of sea level and not a positive movement of land.

Suess in fact doubted the reality of uplifting. As mentioned previously, he limited the movements of the crust to processes of collapse.
These processes, whether occurring at the bottom of the ocean or
along margins of continents, led respectively to a lowering of the
ocean or to an increase in the sea's area. Thus, by successive phases
of collapse, marine shorelines regressed along continents that remained static.

## Eustatism

The interest of Suess's hypothesis lies in his theory of global movements: the level of *all* oceans is lowered simultaneously. Therefore, long-distance correlations again became possible, whereas in the theory of uplifting, moving shorelines could not be used for dating.

Suess called these movements of the ocean **eustatic.** He forcefully stated that the reality of the role of eustatic movements is beyond any doubt.

However, the history of the sea does not consist of regressions only, but also of alternating transgressions. Suess wrote, "Each eustatic movement increases over the entire globe the slope of streams and incites degradation of emerged lands, hence the greater supply of sediments." Deposition of sediments leads, he said, to a rising of the seafloor and eventually generates a positive eustatic movement, or transgression. Thus, a kind of balance is maintained, although negative effects predominate because of the collapsing crust.[7]

In France the eustatic theory became popular after Charles Depéret (1854–1929) had accepted and used it in a somewhat dogmatic fashion. He related eustatism not to collapse of the crust, but to variations in volume of oceanic waters during alternating glaciations. During glaciations, the volume of water is reduced, being stored as ice on continents; during warming periods, the water of glaciers is returned to the ocean. Convinced of the stability of emerged lands, Depéret searched for marine terraces along coastlines and measured their elevation to establish their age.[8] The **altimeter** became a dating instrument for the Pleistocene geologist.

Neither Depéret's theory nor Élie de Beaumont's survived its author. These theories were oversimplified and kept alive only by the authority and prestige of their respective authors.

## The Nappes

Meanwhile, Suess was working on volume 3 of *Das Antlitz der Erde*, which he completed in 1909. Very early, he had understood the role of lateral movements and, in turn, later benefited from the views Marcel Bertrand derived from Suess's earlier work. This interchange of knowledge resulted in Bertrand's theory of nappes (thrusted recumbent folds).

Bertrand's death in 1907 put an end to this cross-pollination.

However, between 1887 and 1900, his works were of exceptional quality. After his paper on the structure of the Alps, he published in 1887 a work on the **klippe** of Beausset. In 1890 he prepared a synthesis of his ideas in a book to be entitled *Mémoire sur les refoulements qui ont plissé l'écorce terrestre et sur le rôle des déplacements horizontaux* (On the overthrusts which have folded the terrestrial crust and on the role of horizontal displacements).[9] Professional scruples made him postpone its publication in order to introduce corrections, and because his subsequent observations led to further changes, the memoir was published only posthumously in 1908.

## The Klippe of Beausset

The most famous overthrust studied by Bertrand is the one of the Beausset Basin [France, Département du Var]. Masses of Triassic rocks rise above Cretaceous rocks containing **lignite** in an abnormal stratigraphic position.

P. Termier wrote:

Marcel Bertrand worked all winter 1887 on his talk of March 21. Spring took him back to Provence . . . and in May 1887 he found the explanation for the *stratigraphic anomaly* of Beausset, which, since geological exploration had begun in Provence and since the little mine of Cadière was operated, represented an unnerving enigma.* The enigma was solved and all difficulties vanished if one assumed that the Triassic rests on the Cretaceous as an *erosional remnant of an overthrusted recumbent fold,* which came from the south and extended further north. Gradually, Marcel Bertrand reached the inevitable conclusion: Provence is a land of *nappes* similar to those found in the French-Belgian coal basin and in the Alps of Glaris . . . ; recumbent folds *thrusted* from south to north for *several kilometers* over their substratum.[†]

---

* Bertrand wrote as follows on the former interpretation: "For geologists who have successively studied the region, the explanation was obvious: this Triassic was a reef forming a relief on the sea floor and the Cretaceous sediments were deposited onlapping the flanks of this small island, in the same position shown today by their horizontal beds" (Marcel Bertrand, *Mémoire sur les refoulements qui ont plissé l'écorce terrestre et sur le rôle des déplacements horizontaux* [Paris: Gauthier-Villars, 1908], chap. 15).

† Pierre Termier, *A la gloire de la terre* (Paris: Desclée de Brouwer, 1922), 156–158.

Bertrand's students, together with other French geologists, took an active part in the study of nappes. His successor at the School of Mines was Pierre Termier (1859–1930). With his synthetical views on the Alps and on their extension across the Mediterranean Sea, he was instrumental to the understanding of the structural unity of the chain. Termier's eloquence turned many into geologists, although critics called him a "geopoet" because of his lyrical style. He apparently did not mind. He wrote a series of books popularizing geology, among which we remember vividly *A la gloire de la terre* (In praise of the earth).[10]

## Émile Haug

Two years younger than Pierre Termier, Émile Haug (1861–1927) deserves particular attention. After earlier studies on ammonites at the University of Strasbourg, he left for Paris and became interested in Alpine tectonics. His memoir, whose title translates in English to "Geosynclines and Continental Areas: On Transgressions and Regressions," represents a milestone in the history of structural geology.[11] He wrote that **geosynclines** are marine troughs in which sedimentation is continuous for long periods of time because their constant sinking (**subsidence**) prevents them from being filled. They play an important role in the history of the earth because, upon uplifting, they become mountain chains.

The first presentation of this concept was by James Dana (1813–1895), who was also the author of a very popular *Manual of Geology*.[12] In a series of articles published in 1873, Dana developed a theory of mountain building and of the origin of continents and oceans in which he discussed the problems of subsidence, uplifting, deformation, and metamorphism, all essential aspects of orogenies. He criticized the ideas of the American geologist James Hall (1811–1898, not to be confused with James Hall from Scotland, often called Sir James Hall), who had studied the Appalachians. Before Dana, James Hall had observed that the same beds displayed a much greater thickness in the Appalachians than in adjacent areas. He proposed the concept of accumulation of sediments in troughs.[13] Dana accepted these ideas in part, though he criticized Hall for not explaining mountain building.

In 1900 Haug restudied the question and proposed a distinction that was soon widely accepted. According to him, the Alps show thick and continuous Mesozoic rocks, which contrast with the thinner contemporaneous beds of the Paris Basin (continental area). These beds display a certain discontinuity due to alternating transgressions and regressions.

Haug's *Traité de géologie* (Treatise of geology) was published between 1907 and 1911 and can be compared to a certain extent with Suess's *Das Antlitz der Erde*, although Haug's style was more didactic and less descriptive.[14] Suess died in 1914. In the last volume of his great work, he stated that the earth consisted of successive envelopes: the *nife*, a core of nickel and iron 5,000 kilometers thick; the *sima*, a cover of silica and magnesium about 1,500 kilometers thick; and the *sal*, a thin external layer of silica and aluminum.

These great syntheses in structural geology were but one aspect of the scientific activity in the earth sciences at the beginning of the twentieth century. Other fields developed at the same time, in particular, petrography, mostly because of the generalized usage of the petrographic microscope and the improvement in techniques of preparing thin sections of rocks. Henry C. Sorby (1826–1908) established the fundamentals of modern sedimentary petrography and sedimentology.[15] Ferdinand Fouqué (1828–1904) and Auguste Michel-Lévy (1844–1901) used chemical data together with mineralogical composition and texture to classify igneous rocks. Their well-known table of classification became the base of modern classifications.[16]

Despite the importance of these investigations, which demonstrate how progress in technology contributed to a steady advancement in petrography and chemistry, I need to hurry on. Indeed, a revolution was in the making in tectonics. A German meteorologist, Alfred Wegener, presented the crazy idea that continents of sal were drifting on a substratum of sima. The general acceptance of his ideas took a long time.

# Chapter 16

# Continental Drift

## Alfred Wegener

On January 6, 1912, Alfred Wegener (1880–1930) read a paper at the Geological Association at Frankfurt am Main entitled, "Die Herausbildung der Grossformen der Erdrinde (Kontinente und Ozeane) auf geophysikalischer Grundlage" (Geophysical basis of the evolution of large-scale features of the earth's crust).[1] Wegener's fundamental idea was that continents were joined together at a certain time in the past; thereafter, they drifted like rafts over the ocean floor, finally reaching their present position. This revolutionary idea was published in a book in 1915, *Die Entstehung der Kontinente und Ozeane* (The origin of continents and oceans).[2] Three revised editions followed after World War I, in 1920, 1922, and 1929, each containing new data.

Interested in meteorology, Wegener had joined a Danish expedition to northeastern Greenland in 1906–1908. Between 1908 and 1912, he taught meteorology at the Physical Institute in Marburg, Germany. Together with Captain J. Koch, he led the second expedition to Greenland in 1912–1913, a trip that allowed him to cross the ice cap of Greenland. In 1929 Wegener organized a third expedition to the western coast of Greenland in preparation for an important trip scheduled for the following year. He planned to study an area across the ice cap along the 17° parallel on a route slightly to the south of the 1912 expedition. In 1930–1931 he tried to set up three stations: a station in the west that had been explored in 1929; one in the east; and one in the middle of the ice cap, at 3,000 meters elevation and 400 kilometers from the western station.

The fourth expedition was Wegener's last one. Bad weather delayed the setting up of the three stations. On August 6, 1930, he wrote, "August shall bring the decisive battle, particularly in regard to the middle station on the ice."[3] On September 21 he left the western station to bring supplies to the middle station. But the trip took longer than planned, and he reached the station only on October 30, without any supplies. He left the next day—it was his fiftieth birthday—with a young man from Greenland to return to the coast. His body was found under the snow on May 8 of the following year, wrapped in his sleeping bag and a reindeer hide, halfway between the two stations. His hands showed no frostbite, which seemed to indicate that he had not died while on the road from the cold, but in his tent from cardiac arrest due to excessive physical effort.[4]

Wegener's ideas were debated during his lifetime, and acceptance came only in the 1960s. For thirty years, the theory met with an incredible amount of skepticism both in Europe and in North America.[5]

## Continents as Jigsaw Fits

Wegener wrote:

> The first concept of continental drift first came to me as far back as 1910, when considering the map of the world, under the direct impression produced by the congruence of the coastlines on either side of the Atlantic. At first I did not pay attention to the idea because I regarded it as improbable. In the fall of 1911, I came quite accidentally upon a synoptic report in which I learned for the first time of palaeontological evidence for a former land bridge between Brazil and Africa. As a result I undertook a cursory examination of relevant research in the fields of geology and palaeontology, and this provided immediately such weighty corroboration that a conviction of the fundamental soundness of the idea took root in my mind.[6]

The correspondence in shape between the coasts of South America and Africa, which makes the Atlantic Ocean resemble a huge valley with parallel flanks, had impressed earlier authors. In 1858 Antonio Snider-Pellegrini (active 1851–1861, dates unavailable) published an illustration of the world before and after the separation of

Africa and South America in *La Création et ses Mystères dévoilés* (The creation and its mysteries explained).[7] Although some textbooks still claim Francis Bacon as a forerunner of drift, he compared only the similar triangular shapes of South America and Africa, not their corresponding coasts.[8]

Paleontological and geological arguments demonstrated correlations between the two continents. Marcel Bertrand and Eduard Suess had stressed structural analogies between Paleozoic chains on both sides of the Atlantic, that is, between the Appalachians and the Hercynian chain. Suess coined the word **Gondwana** for a continent joining Africa and India across the Indian Ocean and the island of Madagascar. Since the same flora was found in Carboniferous rocks in these countries, including Brazil, the continent of Gondwana was believed to have included all the continents in the Southern Hemisphere. In addition, a fossil plant called *Glossopteris* was found on all these southern continents, so the name *Glossopteris* flora was given to all plants of the end of the Paleozoic found on the former continents of Gondwana. A small Permian reptile called *Mesosaurus* was further proof of a former Gondwana: it was found only in Africa and South America, not in Eurasia nor North America.

Similar correspondences were observed in different regions of different age. In the last edition of his book, Wegener referred to studies on the garden snail, distributed "from Southern Germany via the British Isles, Iceland and Greenland across to the American side."[9] Finally, correlations were established between continents of the Southern and the Northern Hemispheres.

Structural geology also revealed connections across the Atlantic. Besides the examples of the Hercynian chain and the Appalachians mentioned by Bertrand, Wegener showed that the South African chains of the Cape of Good Hope extended to the region of Buenos Aires, Argentina. Similarly, the gneissic plateau of Africa corresponds to that in Brazil.

## Land Bridges or Drifting?

There were two different ways to explain these continental jigsaw fits. Suess held that the crust is continuously collapsing. That gave an easy explanation for the fits: the former continents had been larger than today, and their fragments are now resting at the bottom of the ocean. Early twentieth-century authors believed the existence of former land bridges between present-day continents explained fauna

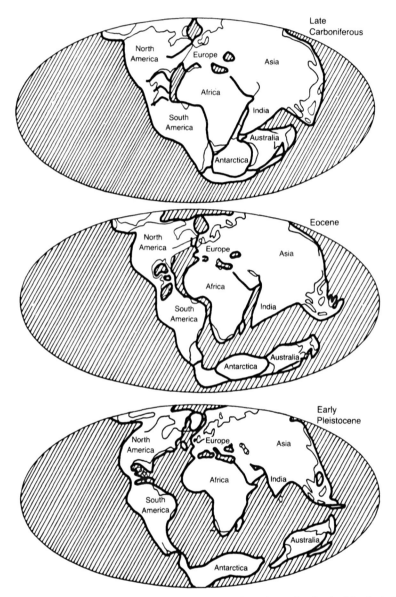

*Figure 16.1.* Reconstruction of the Globe during Three Geological Periods According to the Theory of Continental Drift *(From Wegener,* Die Entstehung der
Kontinente und Ozeane, *1912). In the Late Carboniferous, continents were
welded into a single mass. In the Eocene, the Atlantic and the Indian oceans
began to open. In early Pleistocene, only an isthmus between South America
and Antarctica remained. Arabia and Africa were joined before the opening of
the Red Sea.*

and flora from one continent to another. Numerous imaginary bridges were proposed: Africa and Brazil were connected by the bridge called Archhelenis, and Europe and North America by Archatlantide. Smaller land bridges across the Indian Ocean joined Madagascar to India and India itself to Australia.

To these explanations of land bridges, Wegener opposed the idea of continental drift, or simply drift as it was soon called. Instead of huge land bridges across oceans, he proposed that continents were originally joined together and then gradually separated (fig. 16.1). He wrote: "South America must have lain alongside Africa and formed a unified block which was split in two in the Cretaceous; the two parts must then have become increasingly separated over a period of millions of years like pieces of a cracked ice floe in water."[10]

Wegener did not present his theory as a fantasy of floating and breaking icebergs. On the contrary, each argument was based on the latest data. True, certain correlations of fauna or flora could have been equally well explained by land bridges, but the theory of collapsed land bridges was often faced with simple objections. For instance, in the case of the garden snail, Wegener objected: "Even if we neglect the fact that the theory of sunken continents is untenable on geophysical grounds, this explanation is still given by drift theory, because it must interpolate a very long hypothetical bridge in order to connect the two small areas of distribution. . . . Some bridges even stretched across different climatic zones. It is therefore certain that the bridges could not have been used by all the animals on the continents that they connected."[11]

## The Alpine Shortening

Wegener argued that the theory of collapse was conceivable only in Suess's day, when the contraction of the globe was generally accepted. This theory explained mountain building as a kind of wrinkling because naturalists were not aware of the magnitude of tangential movements during mountain building. Believers in the contraction theory reduced the Alps by numerous overthrusts to one fourth, if not one eighth, of their original width. Since the present width of the Alps is about 150 kilometers, the original width must have been 600 to 1,200 kilometers and must have covered five to ten degrees latitude. R. Staub's estimates were even higher: in 1924 he postulated an Alpine compression of 1,500 kilometers.[12] Staub, as

quoted by Wegener, said that "Africa must have been displaced relative to Europe by this amount. What is involved here is a true continental drift of the African mass and an extensive one at that."[13]

Wegener claimed that the discovery of radioactivity completed
the destruction of the old theory of secular cooling. Radioactive elements within the crust prevent its cooling; hence, "it is no longer
possible, as it once was, to consider the thermal state of the earth as a
temporary phase in the cooling process of a ball that was formerly at
a higher temperature. It should be regarded as a state of equilibrium
between radioactive heat production in the core and thermal loss
into space."[14]

## Isostasy

The theory of isostasy also refuted the concept of collapsed land
bridges. Mentioned earlier (chapter 11), the theory says that the
earth's crust floats in a hydrostatic equilibrium on a denser, viscous
substratum, forming the floor of the oceans. If continents are lighter
than their substratum, they cannot sink to the bottom of the ocean
unless they have been overloaded. Hence, the theory of collapsing
land bridges is incompatible with isostasy.

The concept of isostasy is based on a series of observations made
in the middle of the nineteenth century. In India, measurements of
the meridional arc across the continent revealed a discrepancy
between astronomical and geodetic (triangulation) measurements
between two cities. John Henry Pratt (1800–1871), archdeacon of
Calcutta, interpreted this difference as the effect exerted by the Himalayas on the direction of the plumb line. The plumb bobs, which
under normal conditions point toward the center of gravity of the
earth, were deflected because of the vicinity of attracting masses.
Pratt made the appropriate calculations and found to his surprise
that the Himalayas should have produced an even greater deflection.

Pratt's paper, sent to the Royal Society of London,[15] aroused the
interest of George Biddel Airy (1801–1893), the royal astronomer of
the United Kingdom. He imagined the crust floating on a fluid of a
higher density than the crust and compared it to a raft made of tree
trunks floating on water. He pointed out that the trunks that rise
highest above the surface should also be immersed deeper in the
fluid, following Archimedes' principle whereby the weight of the immersed body is equal to that of the fluid displaced. Airy assumed that
"roots" of the lighter crust are present underneath the Himalayas and

Tibet, extending into the underlying fluid and compensating, to a certain extent, for the deflection produced by the mountain masses. In summary, a mountain with light roots is not heavier than a plain lying directly upon the denser fluid.[16]

In 1859 Pratt refuted Airy's hypothesis, saying that the crust is denser than the underlying fluid because, although both consist of the same material, cooling and contraction makes the crust heavier. Since contraction is less strong in mountain regions than in plains (and, of course, at sea), the density would be lower in mountain ranges. Hence, lower density compensates for higher altitude.[17]

In 1889 C. E. Dutton (1841–1912) called this phenomenon the theory of isostasy.[18] Measurements soon verified the compensating effect of reliefs; that is, mountains do not have a greater gravitational attraction than oceans, despite their greater mass. Explanations varied, however, from one author to another. In the twentieth century, seismology confirmed Airy's ideas and showed that the lower boundary of the crust (recognized by the existence of a surface of **discontinuity**) is deeper below mountains—as if roots indeed existed there. It was also established that the crusts of the continents and of the ocean are of different nature and of different density.

Wegener thus wrote: "The correct interpretation may be found in amalgamation of both concepts. In the case of mountain ranges, we have to do basically with thickening of the light continental crust, in Airy's sense; but when we consider the transition from continental block to ocean floor, it is a matter of difference in type of material, in Pratt's sense."[19]

As a result, Wegener understood that if continents do not have the same composition as oceans, then they are not interchangeable by random collapses of the crust, as Suess assumed. More importantly, isostasy implied vertical movements of the crust. It was known that Scandinavia sank under the weight of the ice during the Pleistocene glaciation, and then bounced back during warmer postglacial times (see chapter 11). Similarly, when continental crusts become thinner by erosion, they become lighter and rise.

## Drift versus Permanence

If vertical movements are possible, asked Wegener, why should that not be the case for horizontal displacements? He assumed that the material forming the ocean floor extended beneath the continents. He equated this material with Suess's sima and, following the work of seismologists, identified it as basalt. He gave the name *sial* to the material of granitic and gneissic composition that forms continents, slightly changing the term *sal* to avoid any confusion with the word *sal* (salt in Latin) proposed by Suess.

To consider sial as a discontinuous layer, limited to continental masses, was a noticeable change from Suess's interpretation of a continuous layer. This made it possible for continents to move horizontally like rafts. Wegener's theory was therefore in agreement with the knowledge of his own time, which disavowed Suess's synthesis made at the end of the previous century.

Some conservative American geologists, however, believed with Bailey Willis (1857–1949) in the permanence of the features of the earth. Willis said, "The great ocean basins are permanent features of the earth's surface and they have existed where they are now, with moderate changes of outline, since the waters first gathered." [20]

A similar idea had been proposed in 1846 by James Dana.[21] He accepted the contraction theory, which postulated that the earth was cooling, but stated that the continents had contracted before the oceans. During the Silurian, contraction of continents led to subsidence of ocean floors. The waters that had initially covered the entire earth were now assembled in ocean basins. Because ocean floors underwent stronger contractions, they exerted lateral pressure upon the continents, which along their borders produced geosynclines and **geanticlines** that surrounded the continents.[22] Greatly influenced by Dana, American geologists never accepted collapses of continents as proposed by Suess. Modern historians maintain that, around 1900, two schools existed: that of Suess and his followers in Europe, who emphasized repeated collapses; and that of permanists, who stressed permanence while accepting contraction. American geologists from Dana to Willis persisted in that belief.[23]

## Paleoclimatic Arguments

Once the difference in composition between oceans and continents was understood, analogies of faunas and floras between distant continents remained to be explained. So did climatic variations through geological time. It was known that during the Late Paleozoic, Europe had a warmer climate, which allowed the development of the Carboniferous flora; whereas land masses of the former continent of Gondwana, in the Southern Hemisphere, display today remains of glaciations and a *Glossopteris* flora of cold climates dating back to the same period. But as long as proofs of glaciations were not common knowledge, it was possible to claim that the globe had gone through warmer periods in the past. Buffon and many other naturalists had assumed that the surface of the earth was gradually cooling. But the *Glossopteris* flora compelled scientists to accept the idea of a distribution of climates different from that known today. More precisely, today's continents must have been in different positions with regard to the poles.

Although the law of permanence was not opposed to the idea of a displacement of the poles, it was imperative to know where the Carboniferous pole was located in order to determine whether today's widely separated land masses—Australia, India, Africa, Brazil, and even Antarctica—could have been covered at the same time by an ice cap. If these continental blocks could be joined into a single one, the problem would be solved.

Wegener could thus claim that his theory was a synthesis of all existing data and explanations. He wrote:

> If drift theory is taken as the basis, we can satisfy all the legitimate requirements of the land-bridge theory and of permanence theory. This now amounts to saying that there were land connections, but formed by contact between blocks now separated, not by intermediate continents which later sank; there is permanence, but of the area of ocean and area of continent as a whole, but not of individual oceans or continents.[24]

Wegener's theory was certainly very appealing. But for his contemporaries, it had one serious flaw: it did not explain the forces or the mechanism that moved the continental rafts. Wegener believed that the drift toward the equator could be explained by a "flight from the poles" (*Polflucht*) or a push toward the equator, and the westward displacement could be attributed to "tidal friction." These forces supposedly not only pushed sialic continents over their sima substratum, but also folded the sediments, which they transported, into mountain chains. Clearly, Wegener's assumed forces were incapable of performing such a task. He recognized modestly that "the Newton of drift theory has not yet appeared."[25]

## Wegener's Opponents

Wegener's theory was thoroughly discussed all over the world. In 1923 the Geological Society of France organized a colloquium at the request of its president, Paul Lemoine (1878–1940). Drift was strongly criticized by Léonce Joleaud (1880–1938).[26] Another debate had taken place in England the year before without any better conclusion.[27] In 1926 a symposium on continental drift was organized by the American Association of Petroleum Geologists. Discussions were heated, but in the end opponents of drift won out, in particular because of the insufficient nature of paleontological and geological arguments and the lack of an adequate mechanism for moving continents—this in spite of the efforts of Waterschoot van der Gracht (1873–1943), who had organized the symposium and who was enthusiastically in favor of continental drift.[28]

The most decisive objections were raised by geophysicists. To assume the gliding of sial over sima, Wegener had to admit the fluidity of sima. He claimed that sima melted at a lower temperature than sial, but experiments proved the opposite. Furthermore, seismic waves demonstrated that the substratum of oceanic depths is rigid. The opponents therefore openly stated that the theory of continental drift was unscientific.

The theory also encountered strong opposition in Germany. The main opponent was Hans Stille (1876–1966), a well-known structural geologist. He subdivided the earth's history into a series of tectonic phases that recall the ideas of Élie de Beaumont (chapter 12). Stille was a **fixist** and hence believed in the permanence of oceans.[29]

## Wegener's Allies

The American scientist F. B. Taylor (1860–1938) was actually a precursor of Wegener. In 1910 he published a long paper demonstrating that the distribution of Cenozoic mountain ranges, particularly in Asia, suggested a displacement of continents from north to south. But he rejected isostatic adjustments as a mechanism for drifting continents. He interpreted the **Mid-Atlantic Ridge** as the trace left by the separation of Africa from the Americas.[30] The impact of Taylor's theory was such that contemporaries often called drift the Taylor-Wegener theory.

In Europe, Wegener was supported by Émile Argand (1879–1940), a renowned Swiss expert of the Alps and of world tectonics. In his address at the International Geological Congress in Brussels (August 10, 1924), which was published in 1922, he assumed that the entire Alpine system, stretching from the western Alps to the Himalayas, originated from a drift of the Gondwana continent against **Eurasia.**[31] He coined the term *mobilism* for ideas that considered horizontal displacements to be the major factor in the formation of nappes. Rudolf Staub (1890–1961), another Swiss structural geologist, held similar views.

Wegener's major supporter was the South African geologist Alexander L. Du Toit (1878–1948). In 1927 he published a long paper comparing the geology of South Africa and South America. Wegener was very impressed and quoted him in the last edition of his book. In 1937 Du Toit published *Our Wandering Continents*, dedicated "To the memory of Alfred Wegener for his distinguished services in connection with the geological interpretation of our Earth."[32] He presented new arguments in favor of the drift theory, showing in particular that not only the Alpine chains, but also earlier orogenic belts (Hercynian and Caledonian) could now be explained by continental drift. Wegener had joined all continents in Late Paleozoic as a single mass, called Pangea, but Du Toit introduced the idea of two more or less independent supercontinents: **Laurasia** in the north, and Gondwana in the south.

Nevertheless, the major obstacle to the acceptance of the continental drift theory was the question, What mechanism is pushing the continents?

## Convection Currents

In an article entitled "Radioactivity and Earth Movements," Arthur Holmes (1890–1965), a well-known Scottish geologist and professor at the University of Edinburgh, proposed in 1928 a motor for the drift.[33] He stated that ordinary volcanic activity was insufficient to discharge the amount of heat generated by radioactivity in the substratum that was believed to rise to the earth's surface. It was therefore necessary to invoke **convection currents** in the substratum.

He wrote that when a liquid is gently heated from underneath, heat diffuses gradually upward until a critical temperature gradient is reached. If the fluid is strongly heated, the gradient is exceeded and at a certain point currents are generated that make the liquid rise and then fall along a pattern of convection circulation. He believed that a similar situation might occur within the fluid substratum underlying the earth's crust.

According to Holmes, granites in continents are particularly rich

## Measuring Continental Drift

As a typical uniformitarian, Wegener hoped to prove the present-day drift of continents by geodetic measurements, saying:

> If continental displacement was operative for so long a time, it is probable that the process is still continuing, and it is just a question of whether the rate of movement is enough to be revealed by our astronomical measurements in a reasonable period of time.
>
> P. F. Jensen carried out new longitude determinations in western Greenland during the summer of 1922 [the earlier ones dated back to 1823] with this in mind, using the far more precise method of radio telegraphy time transmissions. . . . [For this purpose he established] an observatory at Kornok, in the favorable climate of the upper section of Godthaab Fjord. . . . The determination of the longitude of Kornok has now been repeated (summer, 1927) by Lieutenant Sabel-Jörgensen, using the modern impersonal micrometer which eliminates the "personal equation." This allows for greater accuracy to be achieved than was possible in Jensen's measurements.
>
> Comparison with Jensen's figures yields an increase in the longitude difference relative to Greenwich, i.e., in the distance of Greenland from Europe, of about 0.9 seconds (time) in five years, or about a rate of 36 m/yr.*

Unfortunately, these values were erroneous. Measurements of longitude were too imprecise at that time to determine continental drift. A much longer duration was required to establish the rate of separation between Europe and North America, known to be a few centimeters per year.

---

*Alfred Wegener, *The Origin of Continents and Oceans*, trans. John Biram from 4th rev. German edition (New York: Dover, 1966), 23, 27–29.

in radioactive elements, so the temperature beneath them should be higher than under the ocean. Currents would thus rise under continents, spread horizontally toward their peripheral regions, and then move downward when encountering, at the edges of continents, the weaker currents of the oceanic area (see fig. 16.2).

Holmes declared himself in favor of continental drift in his later work, *Principles of Physical Geology* (1964), saying: "What is really important is not to disparage Wegener's great achievement because

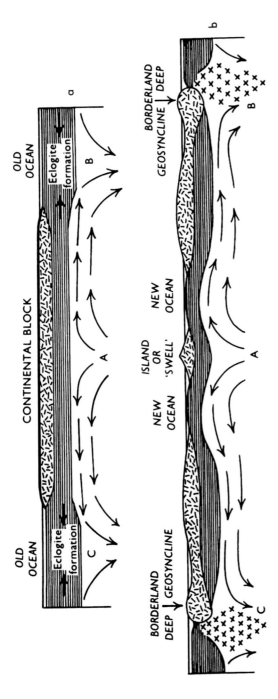

Figure 16.2. Arthur Holmes's Convection Currents Hypothesis (Transactions of the Geological Society of Glasgow 18 [1928–1929]: 579). This hypothesis was used to explain continental drift and the development of new ocean basins when it was thought that the oceanic crust was a thick continuation of the continental basaltic layer (horizontal line shading). (a) A current ascending at A spreads out laterally, extends the continental block, and drags the two main parts aside, provided the obstruction of the old ocean floor can be overcome. This is accomplished by the formation of eclogite at B and C, where subcontinental currents meet suboceanic currents and turn downwards. Being heavy, the eclogite is carried down, thus making room for the continents to advance. (b) The foundering masses of eclogite at B and C share in the main convective circulation and, melting at depth to form basaltic magma, the material rises in ascending currents; e.g., at A, healing the gaps in the disrupted continent and forming new ocean floors (locally with a swell of old sial left behind, as in Iceland). Other, smaller current systems, set in motion by the buoyancy of basaltic magma, descend beneath the continents and feed great floods of plateau basalts, or beneath the "old" (Pacific) ocean floor to feed the outpourings responsible for the volcanic islands and seamounts.

some of his conjectures can be shown wrong, but from the wealth of relevant evidence now available to assess the extent to which continental drift and other lateral displacements of the crust are genuine geological happenings."[34]

Wegener died without having understood the implications of Holmes's ideas, or of similar ones from R. Schwinner, whom he had met in college at Graz.[35] Wegener's theory was rejected by most of his contemporaries in 1930, but became suddenly famous in the 1960s.

# The Birth
## of the Oceans

## The Oceanic Rift

The closer we get to modern science, the more delicate the historian's task: How should we present new discoveries that have not matured sufficiently; in other words, discoveries that have not undergone the ample scrutiny and judgment that earlier scientific findings have? In addition, geology is undergoing a revolution that challenges most of its major concepts; these circumstances may or may not be of help to the historian.

It is, however, not preposterous to state that for the last twenty-five years geology has been dominated by the study of oceans. Richard Field of Princeton University is quoted as having said in the 1930s that he "was convinced that the Earth would not be understood as long as investigations were limited to its emerged third portion [continents]."[1] For quite some time, **bathymetric studies** had revealed a ridge of unknown significance in the middle of the Atlantic. In 1954 the seismologist Jean-Pierre Rothé of Strasbourg showed that the systems of midoceanic ridges extending through various oceans formed an active **seismic belt**.[2] According to Xavier Le Pichon, professor at the Collège de France in Paris, Maurice Ewing, director of Lamont-Doherty Geological Observatory at Columbia University, and his co-worker Bruce Heezen proposed the hypothesis that this belt corresponded to a system of troughs similar to that recognized in the middle of the Mid-Atlantic Ridge. Le Pichon recalled that upon joining Lamont-Doherty in 1959 he was sent on a worldwide search for this puzzling trough, called **rift**, which eventually became the key to the new geology.[3]

The Mid-Atlantic Ridge, discovered a century ago when the first telegraph cables were laid on the seafloor, is 1,000 kilometers wide. It rises 2,000 meters or more above the surrounding abyssal plain, which reaches about 6,000 meters depth. The central part of the ridge, characterized by intense seismic activity, is occupied by a rift about 30 kilometers wide and 2,000 meters deep. Explorations conducted by Lamont-Doherty Geological Observatory traced this structure for more than 60,000 kilometers across all oceans. According to Le Pichon, "It became evident that any well-founded model of the evolution of the Earth could not ignore the existence of the rift system."[4]

## The Conveyor Belt

In 1962 Harry H. Hess of Princeton University explained the role of the rift in a paper, "The History of Ocean Basins," in a multiauthor volume published by the Geological Society of America.[5] Applying the concept of convection currents, he proposed that the ocean floor is continuously being produced in the rift of ridges by rising deep-seated magma, which in turn plunges into deep trenches, stretching along the margins of the Pacific and other oceans. Between the zones of rising and plunging, the ocean floor moves, pushed by the tangential force of convection currents, something like a conveyor belt. Continents are carried along by this movement.

Because the concept of continental drift was now related to the renewal of the ocean floor, the focus of investigations began shifting from continents to oceans. During World War II, Hess was commander of a U.S. Navy ship that criss-crossed the Pacific, undertaking bathymetric surveys. During this research, Hess discovered the flat-topped seamounts for which he coined the name **guyots**. He considered these volcanic cones—rising from the seafloor, and whose summits are often today buried in deep water—to have been originally truncated by wave action at sea level and thereafter submerged during the sinking of the ocean floor (subsidence). Hess searched for an explanation and finally came up with the concept of the conveyor belt, which could also account for many other aspects of the behavior of the oceanic crust.

Hess considered his idea highly hypothetical and tried it out on many of his colleagues before publication, thus losing precious time. His paper was therefore preceded by R. S. Dietz's short article proposing views very similar to those of Hess.[6] Although Dietz did not claim to be the inventor of the new theory, he suggested calling it **seafloor spreading**. He was also the first to point out that the sliding surface was not located at the lower boundary of the earth's crust but at a much greater depth.

The reader will remember that Wegener believed in continents of sial floating over sima. Following Andrija Mohorovičić (1890–1936), seismologists recognized a more important boundary, one separating the crust from the underlying mantle.[7] Holmes located the sliding surface at the base of the crust, underneath both continents and oceans (see fig. 16.2), thus answering the objections of Wegener's opponents. Dietz went further and assumed the sliding surface to be at a much greater depth, at the boundary between **lithosphere** and **asthenosphere**.

Changes between Wegener's ideas and the new ones are drastic. Continents no longer drift freely at the surface of a fluid, but are

## The Earth's Concentric Envelopes

The shallow zones of the earth can be subdivided in at least two ways. The first division is made between crust and mantle, the second between lithosphere and asthenosphere.

1. The earth's crust is the most external envelope of the globe. It is a few kilometers thick under the oceans and several tens of kilometers (about thirty) under the continents. The mantle underlies the crust and extends to a depth of 2,900 kilometers. The boundary between crust and mantle is a discontinuity; that is, a sharp increase of the velocity of seismic waves, called Mohorovičić discontinuity after the Yugoslavian seismologist who discovered it in 1909. The oceanic crust consists of basalt, whereas the continental crust contains rocks richer in silica (of the composition of granite). The mantle is assumed to consist of peridotite, a darker rock containing less silica than basalt.

2. The lithosphere (from the Greek *lithos*, stone) includes the crust and the upper mantle. This envelope is about 100 kilometers thick, rigid, and divided into plates. It overlays the asthenosphere (from the Greek *asthenos*, devoid of force), a more fluid layer produced by partial melting of rock materials.

The boundary between lithosphere and asthenosphere was proven by seismology: the velocity of seismic waves decreases slightly when they penetrate the asthenosphere.

enclosed in "plates" extending beneath and around continents under the oceans. Objections of geophysicists to Wegener's theory (see previous chapter) were removed because the asthenosphere has the fluidity the sima lacked. Moreover, like the carapace of a reptile, plates cover the entire surface of the earth. These plates can be renewed on one side only if consumed on the other. The new theory thus implies generation of oceanic crust along midoceanic ridges and consumption of that crust in **oceanic trenches**.

## Fossil Magnetism

**Paleomagnetism**, that is, the reconstruction of the past magnetic field of the earth, played a critical role in the final formulation of the theory of seafloor spreading. Before the publication of Hess's paper, observations were made which revamped the theory of Wegener. These observations showed that lavas produced by volcanoes, in particular basalts, contain magnetic minerals, which orient themselves according to the earth's magnetic field and maintain that orientation upon cooling of the lavas. Once the minerals solidify, the magnetism they possess will remain "frozen" in this position. Hence, the magnetic field that existed on the earth at the time of lava eruption was, so to speak, fossilized. It became possible to reconstruct that particular magnetic field so long as the orientation of the sample was carefully recorded before collection.

Magnetized minerals not only indicate the direction to the poles, they also provide a means of determining the latitude of their origin. The first measurements made in the 1950s indicated that in the past the poles were not at the same location as today. Scientists at Cambridge University showed that the magnetic north pole was during the Precambrian in the middle of the Pacific Ocean, near Hawaii. Subsequently, it migrated westward, reaching Asia, south of Japan, at the end of the Paleozoic, and eventually moving across Siberia to its present location.

Even more interesting was the discovery made when scientists repeated their measurements in North America: the position of the magnetic north pole, at any given time, was different in other continents. The first set of observations made in Europe did not imply a reciprocal displacement of continents because the magnetic poles could have migrated with respect to a fixed terrestrial crust, as was often assumed in the past. The second set of observations made in North America, however, contradicted such an interpretation because the magnetic poles remain stationary at a given time. If their respective magnetic poles were apparently different in Europe and in North America, then in the past the two continents must have had a different position from today. In order to find their *relative* position at a given time, it was necessary to "move" hypothetically one of the continents until the poles of the two sets of measurements coincided. Eureka! Europe and North America were next to one another in the past, just as Wegener had proposed.

Paleomagnetism provided an additional surprise which, at first glance, did not seem related to continental drift but which, through a number of other studies, eventually supported it. The study of relatively recent lavas (less than a few million years old) showed a magnetism close to the present magnetic field; this appeared quite normal. However, in some instances, the orientation was reversed: the south magnetic pole was at the location of the north magnetic pole. At first it seemed preposterous to accept true reversals of the geomagnetic field because self-reversal magnetization of certain rocks exists. The problem changed, however, after completion of thousands of measurements indicating that lavas of the same age displayed magnetic fields of the same orientation but of either normal or reverse polarity. The conclusion was reached that geomagnetic reversals were real but as yet unexplained. The question became far more complex upon discovery that during the last four million years the magnetic field had undergone four reversals and that during each of these episodes additional changes had taken place. Further discussion of this problem is beyond the scope of this book.

## The Decisive Hypothesis

During the latter part of the 1950s, scientists at Scripps Institution of Oceanography criss-crossed the Pacific, recording in continuous profiles the present-day magnetic field. The purpose of this investigation was not to measure **remanent magnetization** in lavas, but to record the intensity of the present-day earth's magnetic field at the surface of the ocean.

When measurements of the intensity of the magnetic field at a given place were compared to calculations made beforehand, an appreciable difference was found, called a magnetic anomaly. This is comparable to a gravity anomaly, which represents the difference at a given place between the calculated pull of gravity and the measured value. Surprisingly, anomalies detected at sea were approximately ten times larger than those previously known on continents. Furthermore, if positive and negative anomalies were represented on a map, they appeared as parallel bands forming a zebra-skin pattern. At first it was thought that anomalies resulted from the disturbing effect on the earth's magnetic field of magnetic masses located beneath the ocean floor, but not exceeding three kilometers of depth.

In reality, these anomalies were the expression of geomagnetic reversals and provided the answer to the question: How is new oceanic crust generated? The decisive hypothesis was proposed in 1963 by Lawrence Morley, Fred Vine, and Drumond Matthews.[8] They assumed that magnetic anomalies resulted from lavas produced in turn during periods of normal magnetic field (positive anomalies) and periods of reversed magnetic field (negative anomalies). In other words, basaltic lavas poured onto the ocean floor on both sides of the rift, cooled, and then moved away from their original position before new lava was extruded. If a reversal in the earth's magnetic field took place between lava outputs, the result would be a series of parallel bands of basalt, each band having minerals with a different magnetic orientation and each band being progressively older as one moves away from the Mid-Atlantic Ridge. These differences might very well account for the regular, parallel zebra-skin pattern.[9]

According to the hypothesis of the conveyor belt, oceanic crust is being continuously generated in the rifts of midoceanic ridges. Therefore, the zebra-skin pattern on both sides of the ridges strongly supports the theory of seafloor spreading because it explains all observations on the distribution of magnetic anomalies. Indeed, the value of a scientific theory lies as much in its retroactive explanation of facts that had remained enigmas as in its prediction of new facts.

## Subduction

The retroactive power of a theory applied also to the problem of the disappearance of older oceanic crust. In 1935 the Japanese geophysicist K. Wadati had shown that the **foci** of earthquakes along the eastern coast of Asia were located on a plane inclined at about 40° that dipped under Asia.[10] The shallowest foci were below the Pacific and the deepest beneath Siberia and China. These observations were repeated by H. Benioff after World War II, but still no explanation seemed plausible.[11]

The theory of the conveyor belt, which requires oceanic crust to be consumed in the same amount as it is generated (to ensure a constant surface), explains the Benioff or Wadati-Benioff zones, which become **subduction** or sinking zones of the lithosphere into the asthenosphere, while rifts become zones of accretion or of generation of the new lithosphere. The term *subduction* was introduced as early as 1951 by the Swiss geologist André Amstutz in his pioneer studies of the structures of the Penninic Alps; he used the term in a sense that is remarkably in agreement with current theories of **plate tectonics**.[12]

Another example of the retroactive character of the new theory is B. Gutenberg's recognition in the 1920s of a "soft layer" at a depth below 100 kilometers where seismic waves moved at low velocity. This layer was reinterpreted in the 1960s as the boundary between lithosphere and asthenosphere. In summary, the theory of seafloor spreading was able to synthesize scattered observations, some of which had remained unexplained for a long time.[13]

## The Synthesis: Plate Tectonics

The keystone to the theory of plate tectonics remained to be put into place. In 1965 Harry H. Hess was at Cambridge University on sabbatical leave from Princeton. The geology department was directed by Edward Bullard, who in 1956 provided one component to Hess's theory. He showed that heat flow along midoceanic ridges was greater than at any other place on earth. Vine and Matthews, also at Cambridge, had just presented their hypothesis (mentioned above) on magnetic anomalies. Tuzo Wilson, who had supported Hess's theory all along, was also at Cambridge; he was the first to use the term *plates* (in 1965), and presented a model of their behavior in terms of plane geometry.[14]

The final act was put together by three young scientists, Jason Morgan, Dan McKenzie, and Xavier Le Pichon, representing the three main institutions—Princeton, Cambridge, and Lamont—that had provided the most important components of the theory. Like Hess, Jason Morgan was from Princeton. On April 19, 1967, he showed in an oral presentation to the American Geophysical Union that all the investigated phenomena led to the concept that the surface of the earth consists of rigid plates that move with respect to the others. He explained the geometry of the plates in terms of a spherical surface rather than a plane.[15] Dan McKenzie from Cambridge

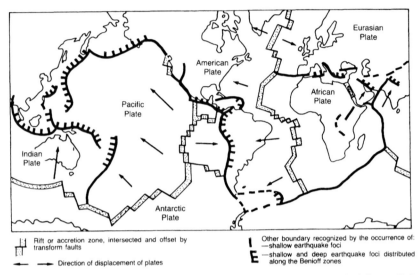

Rift or accretion zone, intersected and offset by transform faults

→ ← Direction of displacement of plates

I Other boundary recognized by the occurrence of:
—shallow earthquake foci

E —shallow and deep earthquake foci distributed along the Benioff zones

*Figure 17.1. Boundaries of the Six Lithospheric Plates (Modified from X. Le Pichon, "Sea-floor Spreading and Continental Drift," Journal of Geophysical Research 73 (1968): 3675, fig. 6. Copyright © by the American Geophysical Union, used with the author's permission).*

University, working at Scripps Institution of Oceanography, contributed to the new ideas in September of the same year in a paper presenting similar ideas.[16] Finally, Xavier Le Pichon, who had been at Lamont since 1959, presented the final synthesis with a sketch of only six large lithospheric plates (compared to the dozen or so Morgan had visualized). His article appeared in June 1968, two months after that of Morgan (see fig. 17.1).[17]

## The Writing of History

This history of geology is nearing its end. Future studies belong to science and not to history. The reader will have noticed that the untangling of each scientist's contribution to plate tectonics is clearly a difficult task. Many names were cited in these few pages alone. Many others deserved to be mentioned. It is evident that it is easier to remember only a few figures for past periods because the collective memory has forgotten the great majority of the contributors. Past centuries seem to render history less difficult because our ignorance unjustly preserved only a few luminaries whose reputations survived the effects of "erosion" better than the rest.

How many authors quoted in this chapter will be remembered fifty or a hundred years from now? It is difficult to answer this question without running the risk of being blamed later on. Oversight has already taken its toll. For instance, Lawrence Morley, who slightly preceded Vine and Matthews, had his article rejected by *Nature* so that it ultimately appeared after theirs. Consequently, his name is omitted in general presentations, which attribute the hypothesis solely to Vine and Matthews. It is true that history might have been written differently had the many protagonists of these discoveries published in a multiauthor work instead of presenting their own versions independently.

When trying to give to all authors their fair share, I perhaps run the risk of citing some whose contributions were modest and inconclusive. Historians who study in detail secondary authors of a given period are well aware of the problem. Are they not often tempted to exaggerate the contributions of these authors and become hero worshippers? Any historical approach is by nature subjective and often biased.

Furthermore, historical "clearinghouses" are often unfair. History is written by the winners, by powerful institutions or cliques rather than individuals. Moreover, time wipes out original and penetrating works to the benefit of a few trailblazing ones, which remain isolated. However, time also has a regulating effect by revealing the turning points in scientific thinking.

Analysis of contemporary science is just as tricky. Its turning points are still unclear, and the direction of its development is even more obscure. Nevertheless, this book cannot reach a final conclusion without indulging in this kind of speculation, as uncertain as it might be. For instance, I have emphasized the recurrence of certain fundamental ideas opposing each other, such as continuity versus discontinuity and irreversible versus reversible (cyclic) change. The debate between uniformitarians and catastrophists rested on these two issues. It is still alive, although most scientists are aware today that uniformitarian and catastrophist concepts interact in most geological processes.

## Continuous Movement

Plate tectonics clearly supports continuity. Earthquake foci along Benioff zones express the repeated episodes of sinking of the lithosphere, while contemporaneous movements open the lips of rifts. Indeed, the tectonic activity of the earth goes on forever.

In the nineteenth century, Élie de Beaumont had already stated that a continuous process (secular cooling) generated discontinuous "revolutions." However, the relationship between these processes remained problematic, so in the end he favored a continuous mechanism that deformed the crust in a uniform manner and a discontinuous process that suddenly accelerated the movement.

Today, more or less continuous movements of the crust are established. For instance, southern Calabria has risen about 1,500 meters in 1.5 million years, at an average rate of one millimeter per year. Such examples of "neotectonics" (present-day tectonic movements) might convert some geologists to a concept of continuity using seafloor spreading as the mechanism, which occurs at the average rate of a few centimeters per year.

## Discontinuity in Many Geological Fields

A certain amount of discontinuity is intrinsic to stratigraphic methodology. If discontinuous variations of faunas did not occur, long-distance correlations would lack precision. After Darwin, biological evolution was assumed to be gradual, but in 1970 a new school of paleontologists took a position against gradualism, stating that evolution is discontinuous. According to the theory of "punctuated equilibria" of Niles Eldredge and Stephen Jay Gould, each species remains in equilibrium for a certain length of time, and then undergoes a sudden change.[18] However, this theory satisfies stratigraphers only if the rupture of equilibrium had a sufficiently global impact to ensure a simultaneous change from one continent to another and hence be of interest for synchronous tectonic phases.

In sedimentology, discontinuity and the concept of the high-energy "rare event" are now widely accepted. Studies of deposits produced by recent tropical hurricanes (**tempestites**) and of earthquake-triggered **tsunamis** have provided textural data that allow geologists to identify similar beds in the geological record and to map their global distribution as a function of plate tectonics.[19] Rare events thus interfere with cyclic sedimentation and show that catastrophes interact with uniformitarian sedimentary processes.

Discontinuity occurs also in structural geology. In 1939, Hans Stille was a powerful supporter of the concept (see chapter 16). He believed that global tectonic "phases" occur synchronously all over the world at more or less regular intervals, phases separated by periods of tectonic quiescence. Each phase lasted about 300,000 years. He created the presently used terminology of tectonic phases, which he named after the region where they are best displayed, such as **Taconic** phase or **Laramide** phase (see appendix). These designations indicate that these global events had more accentuated effects locally.[20]

## Continuous Mountain Building

Stille's phases punctuated the earth's history into periods. In that respect, his phases were a modern equivalent of the systems of Élie de Beaumont in number and in relative suddenness. However, between these two authors came Suess and Bertrand, who divided the earth's history into much longer cycles. Each cycle included continuous processes, and each cycle built a mountain range. In summary, the unit was the chain or, as it is called today, the orogene or **orogenic belt**; and the history of the earth since Early Paleozoic no longer consisted of about twenty phases but of three orogenes: Caledonian, Hercynian, and Alpine.

The concept of discontinuity had lost its former meaning. Although each orogeny itself consisted of tectonic phases, each one was separated from the following by another type of discontinuity. At a certain time the movements of the crust, which for about 200 million years had folded and uplifted a segment of the earth's crust, changed location and became active in another part, leaving the former one to the action of erosion and to oscillations of epicontinental seas. The Hercynian chain during the Mesozoic is a typical example.

Plate tectonics attributed the opening of new oceans to tectonic stresses. Can we not assume that new rifts become active whenever stresses change their place of action? The fundamental discontinuity of the earth's history would thus consist of changes of certain rift systems. Periodically, the lips of rifts would be sealed while others would open. If these revolutions were equated to the three orogenic cycles, there would be three cycles for the 600 million years separating the Present from the end of the Precambrian, hence the figure of 200 million years mentioned above.

Structural geology, conceived in the framework of plate tectonics, shows that mountain ranges are built along the boundary of two plates; but several modes of collision occur, depending on whether plates carry continents or not. A continent consists of light material that cannot sink into the asthenosphere. It may be called "the foam of the earth," as suggested by Claude Allègre.[21] An entirely oceanic plate such as the Pacific plate can sink under the Asian continent along the well-known oblique zone discovered by Wadati and Benioff. When two continental plates collide, as happened between India and Asia or between Africa and Europe, high mountains like the Himalayas and the Alps are formed. The collision of these gigantic blocks generates highly complex structures in which the concepts of continuity and discontinuity are associated. However, this answer is not satisfactory because it is overgeneralized and needs to be examined in detail. Plate tectonics represents a synthesis of many problems that until a few years ago remained unsolved, but plate tectonics does not provide the key to mountain building.

## A Lost Past

Wegener visualized a supercontinent breaking up at the end of the Paleozoic, but said nothing about earlier times. His drift pertained to about 250 million years, that is, to no more than one-twentieth of the earth's history. His opponents asked him to explain older mountain ranges, but data were not available at the time. Today it is generally accepted that continental masses of variable sizes and shapes have always drifted as a function of changing rift systems, and that they were separated and joined repeatedly, as shown by recent detailed reconstructions of Paleozoic paleogeographies.[22] Their history is therefore irreversible, and Wegener's supercontinent (Pangea) at the end of the Paleozoic was only an unusual and fleeting episode in an otherwise widespread global dispersal of continental masses during the earth's history (figs. 17.2 and 17.3).

What is the basic lesson of plate tectonics? We have learned that seafloor spreading introduced a positive component to the history of the earth: new ocean floors are being created. But the theory is also related to the concept of subduction, according to which ancient oceans were consumed and thus disappeared forever. This is indeed the basic lesson of plate tectonics: what is gained in midoceanic ridges is lost in subduction zones. Present-day oceans are young. For instance, during the Jurassic the Atlantic Ocean was no more than a narrow depression in the then-joined continents of Africa and South

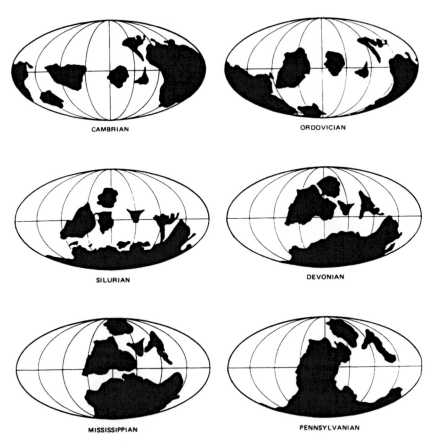

*Figure 17.2. Relative Positions of Drifting Continental Masses during the Time Interval Cambrian through Pennsylvanian (From Don L. Eicher and Lee McAlester,* The History of the Earth, *1980, frontispiece. Reprinted by permission of Prentice-Hall).*

America, like the present-day Afar depression (512 feet below sea level) trending southwest from the area of Djibouti on the Gulf of Aden into the African continent. Subsequently, it resembled the Red Sea when waters invaded it during its widening. Today, its opposite shores are 6,000 kilometers apart. But this young ocean was preceded by others forever lost in the depths of the asthenosphere. In summary, the new geology discovered the oceans, but it also taught that these oceans do not hold the key to the past because former oceanic archives were destroyed by subduction. This key can only have been preserved on continents or not at all.

For our understanding of physical and biological planetary events,

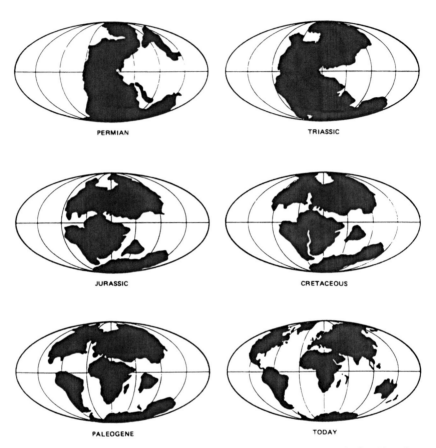

Figure 17.3. Relative Positions of Drifting Continental Masses during the Time Interval Permian through Present (From Don L. Eicher and Lee McAlester, The History of the Earth, 1980, frontispiece. Reprinted by permission of Prentice-Hall).

the destruction of oceanic archives by subduction is certainly a real loss since oceans always occupied three-quarters of the earth's surface. Only very few remnants of oceanic sediments have been preserved in mountain ranges. We should, however, point out that although oceanic archives were certainly thick layers of mud, they were not rich in fossils in comparison with the relatively thin but very highly fossiliferous shallow water deposits laid down on continental shelves and marine embayments by numerous marine transgressions during the **Phanerozoic**. The record of shallow water sediments represents, therefore, the only key to the geological past of the earth, at least before the Cretaceous.

## The Geologist of Tomorrow

The most fascinating aspect of science is that each answer raises new questions. This is particularly true for geology today as it undergoes a major revolution that is changing many of its major concepts. This revolution is far from being completed. However, Le Pichon exaggerated a little when he stated, together with Tuzo Wilson, that this revolution is only an extension of the Copernican revolution.[23] In reality, geology was born and developed in the period after Copernicus. The present-day revolution testifies to the superiority of geophysical methods over those of other "classical" branches of geology. Indeed, the next focus of investigation is to use such methods to discover the motor responsible for the movement of lithospheric plates, and consequently to understand how rift systems changed position during geological time. But the necessity of interaction between the various branches of geology remains of fundamental importance.

Let us analyze a typical example of modern interdisciplinary geological methods. For many years, geologists noticed that the end of the Mesozoic corresponded to a drastic renewal of fauna and flora, with, in particular, the sudden and simultaneous disappearance of dinosaurs on continents and ammonites in the oceans. Hypotheses to explain this global event are of current interest.

Because the end of the Mesozoic was a time of major marine regression, at first only incomplete sequences were available. For instance, in the Paris Basin the sea withdrew at the end of the Cretaceous and returned only in Early Eocene, after a span of time which is difficult to estimate. But several sedimentary sequences were recently discovered in Northern Italy, Denmark, and on the bottom of the Atlantic Ocean, which display uninterrupted sedimentation between Mesozoic and Cenozoic times. These continuous sequences allowed paleontologists to measure the rate of renewal of species. Although the magnitude of faunal changes varied according to environments and to zoological groups, and although the changes were not perfectly simultaneous, it became clear that an enormous biological gap indeed occurred at the boundary of the Mesozoic and the Cenozoic. The cause remains controversial, but abundant data at least allowed geologists and biologists to quantify the problem and to demonstrate its reality.

Geochemistry, combined with detailed paleontology, provided

new markers of this discontinuity. Luis W. Alvarez and his co-workers discovered that the sediments at the transition from Mesozoic to Cenozoic contain an abnormally high amount of **iridium**, a rare trace metal on earth.[24] Because iridium occurs in asteroids and meteorites, its unusual content at the Cretaceous-Cenozoic interface has oriented research toward an extraterrestrial explanation for the paleontological catastrophe: the impact on earth of a comet, asteroid, or large meteorite.[25]

It has become generally accepted that comets or meteorites may indeed be agents of mass destruction on earth even if the impacting body is only ten kilometers in diameter. If the impact occurred on land, it could be compared to a nuclear explosion, vaporizing rocks, generating earthquakes, and ejecting melted and pulverized rock particles that darken the skies; if at sea, the impact would generate gigantic tsunamis. The long-term effects would be even more drastic: global dust clouds, triggering a period of darkness and cold; global wildfires; acid rain; and possibly a long-term **greenhouse effect**. This scenario does not belong to science fiction because tiny metallic spherulites of meteoritic origin and layers of soot have recently been found on a worldwide basis in sediments of the Cretaceous-Cenozoic boundary. Moreover, this catastrophic event may not be as unique as it originally seemed: earlier extinctions in the paleontological record seem to have taken place at intervals of about 26 million years, perhaps indicating a periodicity of the fall of extraterrestrial bodies on earth.[26]

In short, in the fields of structural geology, stratigraphy, paleontology, and sedimentology, that is, in the entire domain of earth sciences, catastrophic events, whether of terrestrial or extraterrestrial origin—perhaps even cyclic with long periodicities—are today considered to have interfered throughout all of geological time in the continuous or short-periodicity movements of our planet's irreversible evolution. Buffon's interplanetary catastrophes and Lyell's uniformitarian concepts are finally considered partners.

The word *geology*, born about two centuries ago, is being gradually replaced by the term *geosciences*. The new term expresses more planetary concerns, since the repository of archives available to earth sciences has enormously increased in recent years. Moreover, the geosciences are interested not only in earth and its satellite, but also in all the planets of our solar system, thus tending to become cosmosciences. Why should earth not show the effects of events that occurred in the history of the planets that surround us?

# The Time Scale of Earth History and the Major Orogenic Phases

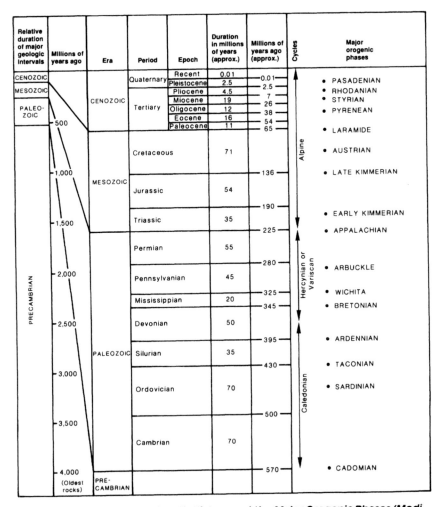

Table 1. *The Time Scale of Earth History and the Major Orogenic Phases (Modified from Don L. Eicher and Lee McAlester, The History of the Earth, 1980, endpapers. Reprinted by permission of Prentice-Hall).*

# Notes

## Introduction

1. D. R. Dean, "The Word Geology," *Annals of Science* 36 (1979): 35–43.
2. William Whewell, *History of the Inductive Sciences from the Earliest to the Present Time*, 3 vols., 2d ed. (London: J. W. Parker, 1847), 3:527.

## Chapter 1: The Beginnings

1. Gaston Bachelard, *Le Rationalisme appliqué* (Paris: Presses universitaires de France, 1949), 9; and *L'Activité rationaliste de la physique contemporaine* (Paris: Presses universitaires de France, 1951), chap. 1.
2. *The Geography of Strabo*, trans. Horace Leonard Jones, 8 vols. (Cambridge, Mass.: Harvard University Press, Loeb Classical Library, 1960), 1:182.
3. Ovid, *Metamorphoses*, trans. Rolfe Humphries (Bloomington: Indiana University Press, 1955), book 15, lines 262–265.
4. *Geography of Strabo*, 1:181.
5. Ibid., 187.
6. For a modern interpretation of the legend of Atlantis, see Dorothy B. Vitaliano, *Legends of the Earth: Their Geologic Origins* (Bloomington: Indiana University Press, 1973), 218–251.
7. Aristotle, *Meteorologica* (Cambridge, Mass.: Harvard University Press, Loeb Classical Library, 1952), 351a.
8. E. Bréhier, *Chrysippe et l'ancien Stoïcisme*, 2d ed. (Paris: Gordon and Breach, 1971), 151.
9. Seneca, *Natural Questions*, trans. Thomas H. Corcoran, 2 vols. (Cambridge, Mass.: Harvard University Press, Loeb Classical Library, 1971–1972), book 3, chap. 3:1–6.
10. Lucretius, *Of the Nature of Things*, trans. William Ellery Leonard (New York: Dutton, 1957), book 6, lines 678–679.

11. Ibid., book 5, lines 436ff.
12. Plato, *Timaeus*, trans. Francis M. Cornford from *Plato's Cosmology* (London, 1937), 22d.

## Chapter 2: At the Center of the World

1. Thomas S. Kuhn, *The Copernican Revolution* (Cambridge, Mass.: Harvard University Press, 1957), 108.
2. Pierre Duhem, *Le Système du monde: Histoire des doctrines cosmologiques de Platon à Copernic*, 10 vols. (Paris: Hermann, 1913–1959), 8:7–9, 9:374.
3. Ibid., 9:253–255.
4. Ibid., 274.
5. Ibid., 265.
6. Ibid., 271ff.
7. Ibid., 274.
8. Ibid., 296.
9. Ibid.
10. Ibid., 204.
11. Ibid., 311.
12. Ibid., 299.
13. *The Notebooks of Leonardo da Vinci*, compiled and edited from the original manuscripts by Jean Paul Richter, 2 vols. (New York: Dover Reprint, 1970), 2:211.
14. In regard to the belief that petrified fish are the remains of some traveler's meal, see Bernard Palissy, *Discours admirables* . . . , 1580, in *Oeuvres complètes*, avec des notes et une notice historique par P.-A. Cap (Paris: Albert Blanchard, 1961). We used here the translation by Aurèle La Rocque, *The Admirable Discourses of Bernard Palissy* (Urbana: University of Illinois Press, 1957), 48–52. See also Marguerite Carozzi, "Voltaire's Attitude toward Geology," *Archives des Sciences*, Genève, vol. 36 (1983): 130–131, and the appendix to that article, a facsimile reprint of Voltaire's work in Italian entitled *Saggio intorno ai canbiamenti avvenuti su'l globo della terra* (Paris: Prault, 1746, 4; and Buffon, *Histoire naturelle, générale et particulière* . . . , 44 vols. (Paris: Imprimerie Royale, 1749–1804), 1:281–282.
15. Palissy, *Oeuvres complètes* (1961), 278.
16. *Notebooks of Leonardo da Vinci*, 2:182.
17. Ibid., 183.

## Chapter 3: How the Earth Was Formed

1. Pierre Duhem, *To Save the Phenomena: An Essay on the Idea of Physical Theory*, trans. Edmund Doland and Chanjnah Maschler (Chicago: University of Chicago Press, 1969). See in particular chapter 6, "From Osiander's Preface to the Gregorian Reform of the Calendar," 67–111. Duhem's original book was called *Essai sur la notion de théorie physique de Platon à Galilée* (Paris: Herman & Fils, 1908).
2. Letter to J.-L. Guez de Balzac, May 5, 1631, in René Descartes, *Principes de la philosophie* (Paris, 1747), in *Oeuvres philosophiques*, 3 vols. (Paris: Éditions Garnier Frères, vol. 1, 1963; vol. 2, 1967; vol. 3, 1973), 1:292.

3. René Descartes, *Principes de la philosophie* (Paris, 1747), in *Oeuvres philosophiques* (Paris: Éditions Garnier Frères, 1973), vol. 3, pt. 3, § 19.
4. Ibid., § 20 and 23.
5. Ibid., § 48–52.
6. O. Hamelin, *Le Système de Descartes* (Paris: F. Alcan, 1911), 338.
7. Descartes, *Principes*, vol. 3, pt. 3, § 146.
8. This paragraph and the following two paragraphs are from ibid., vol. 3, pt. 4, § 22–43.

## Chapter 4: The Work of God

1. Jacques Roger, "La théorie de la terre au XVIIe siècle," *Revue d'histoire des sciences* 26 (1973): 23–48, see p. 39 for his study of Thomas Burnet.
2. Thomas Burnet, *The sacred Theory of the Earth, in which are set forth the Wisdom of God displayed in the works of the Creation, Salvation, and the Consummation of all Things, until the Destruction of the World by Fire, including the blessed Millenium, or, the Reign of Christ with his Saints upon Earth* (London: T. Kinnersley, 1816). The first edition in Latin was *Telluris theoria sacra . . .* (London: Kettilby, 1681). See the facsimile edition of *The sacred Theory of the Earth*, with an introduction by B. Willey (Carbondale: Southern Illinois University Press, 1965), 71.
3. Archbishop Ussher, *Annales veteris testamenti a prima mundi deducti* (London, 1650–1654), 1.
4. Jean Delumeau, *La Peur en Occident (XIVe–XVIIIe siècles): Une cité assiégée* (Paris: Fayard, 1978), 215.
5. John Evelyn, *The Diary of John Evelyn, from 1641 to 1705*, ed. William Bray (London: W. W. Gibbins, 1890). See also the reprint edition (3 vols.) with an introduction and notes by Austin Dobson (London: Macmillan, 1906).
6. See Bernard Palissy, *Discours admirables . . .* , 1580, in *Oeuvres complètes* (Paris: Albert Blanchard, 1961); and Pierre Perrault, *De l'origine des fontaines* (Paris, 1674), trans. Aurèle La Rocque, *On the Origin of Springs* (New York: Hafner, 1967).
7. John Ray, *Three Physico-Theological Discourses concerning 1. The Primitive Chaos, and Creation of the World; 2. The General Deluge, its Causes and Effects; 3. The Dissolution of the World and future Conflagration . . .* , 2d ed. (London: Samuel Smith, 1693).
8. Noël-Antoine Pluche, *Le Spectacle de la nature . . .* (Paris, 1732–1750). See also translation by Mr. Humphreys, *Spectacle de la nature: or Nature display'd, being discourses of such particulars of natural history as were thought most proper to excite the curiosity and form the minds of youth*, 5 vols. (London: Franklin, 1740–1753).
9. Élie Bertrand, "Essai sur les usages des montagnes avec une lettre sur le Nil" (Zurich: Heidegger, 1754), in *Recueil de divers traités sur l'histoire naturelle de la terre et les fossiles* (Avignon: Louis Chambeau, 1766), 111–222.
10. Louis Bourguet, *Lettres philosophiques sur la formation des sels et des crystaux, avec un mémoire sur la théorie de la terre* (Amsterdam: F. l'Honoré, 1729).
11. Ibid., 181–182.
12. C. E. Engel and C. Vallot, *Les écrivains à la montagne; "Ces monts*

*affreux . . . (1650–1810)"; Extraits des lettres de Thomas Gray* (Paris: Delagrave, 1934), 21.

13. Albrecht von Haller, *The Poems of Baron Haller,* trans. Mrs. Howorth (London: J. Bell, 1794).

14. Gordon L. Davies, *The Earth in Decay: A History of British Geomorphology, 1578–1878* (New York: Elsevier, 1969), 110. See also Marjorie H. Nicolson, *Mountain Gloom and Mountain Glory: The Development of the Aesthetics of the Infinite* (Ithaca, N.Y.: Cornell University Press, 1959).

15. Bertrand, "Essai sur les usages," in *Recueil,* 215.

16. Bourguet, *Lettres philosophiques,* 183.

17. Bernard le Bouvier de Fontenelle, "Sur les empreintes de plantes dans les pierres," *Histoire de l'Académie Royale des Sciences, Paris* (1718): 5.

18. John Woodward, *An Essay toward a Natural History of the Earth; and Terrestrial Bodies, Especially Minerals; as also of the Sea, Rivers, and Springs. With an Account of the Universal Deluge; and of the Effects It Had on the Earth* (London: R. Wilken, 1695), 236–237.

## Chapter 5: The Birth of Science

1. Nicolaus Steno, *De solido intra solidum naturaliter contento dissertationis prodromus* (Florence: ex Typographia sub signo Stellae, 1669). See the English translation: *The Prodromus of Nicolaus Steno's Dissertation concerning a Solid Body Enclosed by Process of Nature within a Solid,* 2d ed., trans. John Garrett Winter with a foreword by William H. Hobbs (New York: Hafner, 1968).

2. Bernard Palissy, *Recepte véritable* in *Oeuvres complètes* (Paris: Albert Blanchard, 1961), 37–38.

3. Martin J. S. Rudwick, *The Meaning of Fossils: Episodes in the History of Paleontology* (London: MacDonald; New York: Elsevier, 1972), 61–63.

4. Steno, *Prodromus,* 17.

5. Agostino Scilla, *La vana speculazione disingannata dal senso* (Naples: Andrea Colicchia, 1670).

6. Robert Hooke, "Lectures and Discourses of Earthquakes and Subterraneous Eruptions . . . ," in *The Posthumous Works of Robert Hooke . . .* (London: Richard Waller, 1705; facsimile ed., New York: Arno, 1978), 288.

7. Ibid., 291.

8. Gottfried Wilhelm von Leibniz, *Protogaea, sive de prima facie telluris et antiquissimae historiae vestigiis in ipsis naturae monumentis dissertatio, ex schedis manuscriptis viri illustris . . .* (Goettingae: Sumptibus I. G. Schmidii, 1749), 41; and Leibniz, "Protogaea . . . ," in *Acta eruditorum* (Leipzig, 1693), 40–42.

9. Steno, *Prodromus,* 28.

10. Ibid., 30–31.

11. Ibid., 30.

12. Athanasius Kircher, *Mundus subterraneus in XII Libros Digestus; quo Divinum Subterrestris Mundi Opificium . . . ,* 2 vols. (Amsterdam, 1664–1665), 1:194.

13. Steno, Prodromus, 69.
14. Ibid.
15. Ibid., 64.
16. Peter Simon Pallas, Observations sur la formation des montagnes & sur les changemens arrivés au globe, pour servir à l'histoire naturelle de M. le Comte de Buffon (St. Petersburg: Acta Academiae Imperialis Scientiarum, 1777); B. Faujas de Saint-Fond, Histoire naturelle de la province du Dauphiné (Grenoble, 1781); Buffon, "Époques de la nature," in Suppléments à l'Histoire naturelle (Paris: Imprimerie Royale, 1778), vol. 5, and in Oeuvres complètes, 44 vols. (Paris: Verdière et Ladrange, 1824−1832); J.L.G. Soulavie, Histoire naturelle de la France méridionale . . . , 7 vols. (Nîmes: C. Belle, 1780−1784).
17. Hooke, "Lectures," 335.
18. A. Cournot, Essai sur les fondements de nos connaissances . . . , 2d ed. (Paris: Hachette, 1912), 460−461, § 313.
19. Ibid., 447, § 392.

## Chapter 6: On Mountain Building

1. Jean-Jacques Rousseau, La Nouvelle Héloïse: Extraits, modern version of the 1761 original with biographical, historical, and literary notes by J.-E. Morel, 2 vols. (Paris: Librarie Larousse, 1937), 1:25.
2. Benoît de Maillet, Telliamed ou Entretiens d'un philosophe indien avec un missionnaire françois sur la diminution de la mer, la formation de la terre, l'origine de l'homme, &c. Mise en ordre sur les Mémoires de feu M. de Maillet par J.A.G. (Amsterdam: L'Honoré Fils, 1748).
3. Benoît de Maillet, Telliamed, or Conversations between an Indian Philosopher and a French Missionary on the Diminution of the Sea . . . , trans. and ed. Albert V. Carozzi (Urbana: University of Illinois Press, 1968), 34−44.
4. Jean-André Deluc, Lettres physiques et morales sur les montagnes et sur l'histoire de la terre et de l'homme, adressées à la reine de la Grande-Bretagne (La Haye: chez de Tune, 1778), 7 and 188.
5. Déodat Dolomieu, "Discours sur l'étude de la géologie, prononcé . . . à l'ouverture de son cours sur le gissement [sic] des Minéraux, commencé en Ventôse de l'an 5," Journal de Physique 45 (1794): 265.
6. Ibid., 263.
7. P. Buache, "Essai de géographie physique où l'on propose des vûes générales sur l'espèce de Charpente du Globe, composée de chaînes de montagnes qui traversent les mers comme les terres . . . ," Mémoires, Académie Royale des Sciences de Paris (1752), 399−416.
8. Lazzaro Moro, De' Crostacei e degli altri marini corpi che si truovano su'monti (Venice: S. Monti, 1740), 426.
9. Ibid., 262−263.
10. Ibid., 212.
11. Anders Celsius, "Anmärkning om vatnets förminskande," Kungliga Svenska Vetenskaps Acadmiens Handligar Stockholm 4 (1743): 33−50.
12. Carl Linnaeus, Oratorio de Telluris habitabilis incremente. . . . (Leyden, 1744). See also his Amoenitates Academicae, seu dissertationes variae . . . . 7 vols. (Leyden and Stockholm, 1749−1769), 2:16.
13. Linnaeus, Oratorio, § 33.

14. Johann Gottlob Lehmann, *Versuch einer Geschichte von Flötz-Gebürgen* . . . (Berlin: Gottlieb August Lange, 1756).
15. Ibid., 111.
16. Ibid., 132ff.
17. Paul Henri Tiry d'Holbach, *Essai d'une histoire naturelle de couches de la terre* . . . , vol. 3 of *Traités de physique, d'histoire naturelle, de minéralogie et de métallurgie; ouvrages traduits de l'Allemand* (Paris: Jean-Thomas Hérissant, 1759).
18. Paul Henri Tiry d'Holbach, *Système de la nature: Ou des loix du monde physique & du monde moral*, trans. M. Mirabaud (London, 1770).
19. Guillaume-François Rouelle, "Cours de chimie ou Leçons de M. Rouelle recueillies pendant les années 1754 et 1755; rédigées en 1756, revues et corrigées en 1757 et 1758," 2 vols. ms. in Bibliothèque Nationale, Paris, n.a.fr.4043–4044).
20. G. Stegagno, *Il Veronese Giovanni Arduino e il suo contributo al progresso della scienza geologica* (Verona: Tipographia operaia, 1929), 6–7. Stegagno refers to Giovanni Arduino, "Due lettere sopra varie sue osservazioni naturali" [Letters addressed to Antonio Vallisnieri contain original definition of Tertiary system], in *Nuova raccolta d'opusculi scientifici e filologici* (Venice, 1760).
21. Horace-Bénédict de Saussure, *Voyages dans les Alpes, précédés d'un essai sur l'histoire naturelle des environs de Genève*, 4 vols. (vol. 1, Neuchâtel: Samuel Fauche, 1779; vol. 2, Geneva: Barde, Manget & Cie, 1786; vols. 3 and 4, Neuchâtel: Louis Fauche-Borel, 1796), vol. 2, § 919. See also Albert V. Carozzi, "Forty Years of Thinking in Front of the Alps: Saussure's (1796) Unpublished Theory of the Earth," *Earth Sciences History* 8 (1989): 123–140.
22. Pierre Bernard Palassou, *Essai sur la minéralogie des Mont-Pyrénées*, 2d ed. (Paris: Didot jeune, 1784), 89.
23. M. A. Lacroix, "L'exploration géologique des Pyrénées par Dolomieu en 1782," *Bulletin de la Société Ramond* (1917–1918): 120–178.
24. Philippe Isidore Picot de Lapeyrouse, "Voyage au Mont-Perdu et Observations sur la nature des crêtes les plus élevées des Pyrénées," *Journal des Mines* 7 (1797–1798): 39–66.
25. L. F. Ramond de Carbonnières, *Voyages au Mont-Perdu et dans la partie adjacente des Hautes-Pyrénées* (Paris: Belin, 1801), 74.
26. See Peter Simon Pallas, *Reise durch verschiedene Provinzen des Russischen Reichs*, 3 vols. (St. Petersburg: Acta Academiae Imperialis Scientiarum, 1771–1776); and Alexander von Humboldt, "Esquisse d'un tableau géologique de l'Amérique méridionale," *Journal de Physique* 53 (1801): 30–60.
27. Maillet, *Telliamed, or Conversations*, 174.

## Chapter 7: Buffon as Historian

1. Buffon, *Histoire naturelle, générale et particulière* . . . , 44 vols. (Paris: Imprimerie Royale, 1749–1804). This is merely the first edition of Buffon's *Histoire naturelle*, which was followed by many others after his death. Furthermore, Buffon's *Oeuvres complètes*, which includes essentially the same material as *Histoire naturelle* but in a different order, was kept in print for decades. The list of the contents of the first

edition is from *Auteurs modernes: 1, Oeuvres philosophiques de Buffon*, ed. Jean Piveteau, vol. 41 of *Corpus Général des Philosophes Français*; publié sous la direction de Raymond Bayer (Paris: Presses universitaires de France, 1954).

2. René-Antoine Ferchault de Réaumur, *Mémoires pour servir à l'histoire des insectes*, 6 vols. (Paris: Imprimerie nationale, 1734–1742).
3. Buffon, *Histoire naturelle*, 1:19.
4. Ibid., 36.
5. Buffon, *Des Époques de la nature*, ed. Gabriel Gohau (Paris: Editions rationalistes, 1972).
6. Buffon, *Histoire naturelle*, 1:99.
7. Ibid., 124.
8. Mircea Eliade, *Le mythe de l'éternel retour* (Paris: Nouvelle Revue Française, 1969), 92.
9. Buffon, "Époques de la nature," in *Suppléments à l'Histoire naturelle* (Paris: Imprimerie Royale, 1778), vol. 5, and in *Oeuvres complètes*, 44 vols. (Paris: Verdière et Ladrange, 1824–1832), 29:1–254.
10. Ibid., 71.
11. Ibid., 170.
12. Jacques Roger, *Buffon: Les Époques de la nature*, 2d ed. (Paris: Mémoires Muséum d'Histoire naturelle, 1988), vol. 10, ser. C, 49.
13. Buffon, "Époques de la nature," 187–188. See also Roger, *Buffon*, 159.
14. M.-J. Hérault de Séchelles, *Oeuvres littéraires et politiques*, ed. Hubert Juin (Lausanne: Éditions Rencontre, 1970). For the visit to Buffon in September 1785, see p. 86.
15. Buffon, *Histoire naturelle*, 4:v–xvi.
16. Buffon, "Époques de la nature," 34.
17. Roger, *Buffon*, 41.
18. Ibid., 40.
19. Nicolas-Antoine Boulanger, "Anecdotes de la nature," ms. in Muséum d'Histoire naturelle, Paris. See John Hampton, *Nicolas-Antoine Boulanger et la Science de son temps* (Geneva: Librairie E. Droz, 1955); Jacques Roger, "Un manuscrit perdu et retrouvé: Les anecdotes de la nature," *Revue des sciences humaines*, Faculté des Lettres (Lille, France), n.s. 71:231–254.
20. Nicolas-Antoine Boulanger, *L'Antiquité dévoilée par ses usages ou Examen critique des principales Opinions, Cérémonies & Institutions religieuses & politiques des différens Peuples de la Terre* (Amsterdam: Marc Michel Rey, 1766). Boulanger's *Traité du Despotisme Oriental*, published in 1761 by d'Holbach, is actually the last chapter of *L'Antiquité dévoilée*.
21. Buffon, "Époques de la nature," 101–102.

## Chapter 8: At the Service of Industry

1. Johann Gottlob Lehmann, *Versuch einer Geschichte von Flötz-Gebürgen . . .* (Berlin: Gottlieb August Lange, 1756), pt. 4, 132–153.
2. George Christian Füchsel, *Historia terrae et maris ex historia Thuringiae per montium descriptionem erecta* (Erfurt and Gotha: Acta Academiae electoralis maguntinae scientiarum utilium qual Erfordia est, 1761), 2:44–254. See also Füchsel, *Entwurf zu der ältesten Erd-und*

*Menschengeschichte* . . . (Frankfurt and Leipzig, 1773), copy kindly provided by COFRHIGEO (Comité français d'histoire de la géologie), Maison de la Géologie, Paris.

3. Christian Keferstein, "Notice sur Füchsel et ses ouvrages," *Journal de Géologie* 1 (1830): 194.

4. Abraham Gottlob Werner, *Von den äusserlichen Kennzeichen der Fossilien* (Leipzig: Siegfried Lebrecht Crusius, 1774). See also translation by Albert V. Carozzi, *On the External Characters of Minerals* (Urbana: University of Illinois Press, 1962).

5. Jean Baptiste Louis de Romé de l'Isle, *Essai de cristallographie, ou Description des figures géométriques, propres à différens corps du règne minéral* . . . (Paris: Didot jeune, 1772); and René-Just Haüy, *Essai d'une théorie sur la structure des crystaux* . . . (Paris: Gogée & Née de la Rochelle, 1784).

6. Jean-François d'Aubuissons de Voisins, *Traité de géognosie* . . . , 2 vols. (Strasbourg and Paris: F. G. Levrault, 1819), 2:10.

7. Abraham Gottlob Werner, *Kurze Klassifikation und Beschreibung der verschiedenen Gebirgsarten* (Dresden: Waltherische Hofbuchhandlung, 1787). See also Werner, *Short Classification and Description of the Various Rocks*, trans. Alexander M. Ospovat (New York: Hafner, 1971), 17, 21, 95.

8. Aubuissons, *Traité de géognosie*, 2:12.

9. Jean-André Deluc, *Lettres sur l'histoire physique de la terre, adressées à M. le professeur Blumenbach, renfermant de nouvelles preuves géologiques et historiques de la mission divine de Moyse* (Paris: chez Nyon, 1798), 73, 61.

10. Nicolas Desmarest, "Extrait d'un mémoire sur la détermination de quelques époques de la nature par les produits des volcans & sur l'usage de ces époques dans l'étude des volcans," *Observations sur la physique*. . . . 13 (1779): 115–126.

11. Jean-Etienne Guettard, "Mémoires sur quelques montagnes de la France qui ont été des volcans," *Mémoires de l'Académie Royale des Sciences, Paris* (1752): 27–59.

12. Nicolas Desmarest, "Mémoires sur l'origine et la nature du basalte à grandes colonnes polygones, déterminées par l'histoire naturelle de cette pierre, observée en Auvergne," in *Mémoires de l'Académie Royale des Sciences, Paris* (1771), 705–775. See also Desmarest, "Mémoire sur le basalte . . . ," in *Mémoires de l'Académie Royale des Sciences, Paris* (1773), 599–670.

13. Jean-François d'Aubuissons de Voisins, "Mémoires sur les volcans et les basaltes de l'Auvergne," *Journal de Physique* 88 (1819): 432–449.

14. J.L.G. Soulavie, *Histoire naturelle de la France méridionale* . . . , 7 vols. (Nîmes: C. Belle, 1780–1784).

15. P.-A. Boissier de la Croix de Sauvages, "Mémoire contenant des observations de Lithologie pour servir à l'Histoire Naturelle du Languedoc & à la théorie de la terre," *Mémoires de l'Académie Royale des Sciences, Paris* (1746): 713–758, (1747): 699–743.

## Chapter 9: Subterranean Fires

1. James Hutton, *Theory of the Earth, with Proofs and Illustrations* (London: Messrs Cadell, Jr., and Edinburgh: William Creech, 1795; reprint,

Weinheim: H. R. Engelmann [J. Cramer] and Wheldon & Wesley, 1959, 1960).

2. John Playfair, *Illustrations of the Huttonian Theory of the Earth* (Edinburgh: W. Creech, 1802).

3. James Hutton, *Dissertatio physico-medica inauguralis de sanguine et circulatione microcosmi* (Leiden, 1749).

4. James Hutton, "Theory of the Earth; or an Investigation of the Laws Observable in the Composition, Dissolution, and Restoration of Land upon the Globe," *Transactions of the Royal Society of Edinburgh* 1, pt. 2 (1788): 216.

5. Frank Dawson Adams, *The Birth and Development of Geological Sciences* (New York: Dover, 1954), 243.

6. James Hutton, *Abstract of a dissertation read in the Royal Society of Edinburgh, upon the seventh of March, and fourth of April 1785, Concerning the System of the Earth, its Duration and Stability* ([Edinburgh], 1785).

7. Hutton, "Theory of the Earth" (1788), 209–304.

8. Hutton, *Theory of the Earth* (1795), chap. 1.

9. E. B. Bailey, *James Hutton: The Founder of Modern Geology* (Amsterdam: Elsevier, 1967).

10. Jean-André Deluc, *Lettres physiques et morales sur l'histoire de la terre et de l'homme, adressées à la reine de la Grande Bretagne*, 5 vols. in 6 (La Haye: chez de Tune, 1779), 4:630.

11. Hutton, *Theory of the Earth* (1795), 1, pt. 2:200.

12. Ibid., 198–200.

13. S. I. Tomkeieff, "Unconformity: An historical Study," *Proceedings, Geological Association of London* 73 (1962): 383–417.

14. Playfair, *Illustrations*, 1802.

15. James Hall, "On the Vertical Position and Convolutions of Certain Strata, and Their Relation with Granite," *Transactions of the Royal Society of Edinburgh* 7 (1815): 79–108.

16. James Hutton, "Observations on Granite," *Transactions of the Royal Society of Edinburgh* 3, pt. 2 (1794): 77–81.

17. James Hutton, *Theory of the Earth, with Proofs and Illustrations*, 4 parts, vol. 3, ed. Sir Archibald Geikie (London: Geological Society, 1899). See also V. A. Eyles, "Note on the Original Publication of Hutton's *Theory of the Earth*, and on the Subsequent Forms in which it was Issued," in "James Hutton—1726–1797, Commemoration of the 150th Anniversary of His Death," *Proceedings of the Royal Society of Edinburgh*, sec. b, vol. 63, pt. 4 (1948–1949): 377–386 (issued separately February 6, 1950).

18. Horace-Bénédict de Saussure, *Voyages dans les Alpes, précédés d'un essai sur l'histoire naturelle des environs de Genève*, 4 vols. (vol. 1, Neuchâtel: Samuel Fauche, 1779; vol. 2, Geneva: Barde, Manget & Cie, 1786; vols. 3 and 4, Neuchâtel: Louis Fauche-Borel, 1796), vol. 2, § 1184, 1933, 1937, 1938. See also Albert V. Carozzi, "La géologie: De l'histoire de la Terre selon le récit de Moïse aux premiers essais sur la structure des Alpes et à la géologie expérimentale," in *Les savants genevois dans l'Europe intellectuelle du XVIIIᵉ au milieu du XIXᵉ siècle*, ed. J. Trembley, 203–265 (Geneva: Journal de Genève, 1987).

19. Albert V. Carozzi, "Forty Years of Thinking in Front of the Alps: Saussure's (1796) Unpublished Theory of the Earth," *Earth Sciences History* 8 (1989): 123–140.

20. Déodat de Dolomieu, "Rapport fait à l'Institut National (par l'auteur) sur ses voyages de l'an V et de l'an VI," *Journal des Mines* 7 (1797–1798): 425.

21. William Whiston, *A New Theory of the Earth, From Its Original to the Consummation of All Things. . . .* (London: Benj. Tooke, 1696).

22. A. Lacroix, *Dolomieu . . . sa correspondance—sa vie aventureuse—sa captivité—ses oeuvres,* 2 vols. (Paris: Librairie Académique Perrin, 1921), 1:213. See 1788 letter to Picot de Lapeyrouse.

23. Louis Bourguet, *Traité des pétrifications* (Paris: Briasson, 1742), 9–22; Benoît de Maillet, *Telliamed ou Entretiens d'un philosophe indien avec un missionnaire françois sur la diminution de la mer, la formation de la terre, l'origine de l'homme, &c. Mise en ordre sur les Mémoires de feu M. de Maillet par J.A.G.* (Amsterdam: L'Honoré Fils, 1748); Benoît de Maillet, *Telliamed, or Conversations between an Indian Philosopher and a French Missionary on the Diminution of the Sea . . . ,* trans. and ed. Albert V. Carozzi (Urbana: University of Illinois Press, 1968), 34–44; François Ellenberger, "A l'aube de la géologie moderne: Henri Gautier (1660–1737)," *Histoire et Nature* 7; 9–10 (1975; 1977).

24. Henri Gautier, *Nouvelles Conjectures sur le Globe de la Terre* (1721), § 7; booklet reprinted with additional text in *La Bibliothèque des Philosophes et des Sçavans* (Paris, 1723), 491–500.

25. Henri Gautier, *Nouvelles conjectures,* booklet reprinted in "A l'aube," by François Ellenberger, 9–10: 117–118.

26. Dolomieu, "Rapport fait," 7 (1797–1798): 400.

27. John Turberville Needham, *Nouvelles Recherches physiques et métaphysiques sur la nature et la religion, avec une nouvelle théorie de la terre, et une mesure de la hauteur des Alpes* (London: Lacombe, 1769). This addition forms the second part of a translation of Lazzaro Spallanzani, *Nouvelles Recherches sur les découvertes microscopiques et la génération des corps organisés . . . , traduit de l'Italien de M. l'Abbé Spallanzani . . . par M. l'Abbé Regley . . . avec des notes, des recherches physiques & métaphysiques sur la Nature & la Religion, & une nouvelle Théorie de la Terre par M. de Needham* (London: Lacombe, 1769).

28. Needham, *Nouvelles Recherches,* 164.

29. Ibid., 177.

30. Jean Jacques Dortous de Mairan, "Mémoire sur la cause générale du Froid en Hiver & de la Chaleur en Été," *Mémoires de l'Académie Royale des Sciences, Paris* (1719): 104–135; and his "Nouvelles recherches sur la cause générale du chaud en Été et du froid en Hiver, en tant qu'elle se lie à la chaleur interne & permanente de la Terre; en supplément & correction du mémoire qui fut donné sur ce sujet dans le Volume de 1719," *Mémoires de l'Académie Royale des Sciences, Paris* (1765): 143–266.

31. James Hall, "Experiments on the Effects of Heat Modified by Compression," *Journal of Natural Philosophy, Chemistry, and the Arts* 9 (1805): 98–197; and his "Account of a Series of Experiments, Shewing the Effects of Compression in Modifying the Action of Heat," which was published in *Journal of Natural Philosophy, Chemistry, and the Arts* 13 (1806): 328–343, 381–405; 14 (1806): 13–22, 113–128, 196–212, 302–318; and in *Transactions of the Royal Society of Edinburgh* 6 (1812): 71–185.

32. Jean Baptiste Lamarck, *Hydrogéologie; ou, Recherches sur l'influence qu'ont les eaux sur la surface du globe terrestre; sur les causes de l'existence du bassin des mers, de son déplacement et de son transport successif sur les différens points de la surface de ce globe; enfin sur les changemens que les corps vivans exercent sur la nature et l'état de cette surface* (Paris: Jean Baptiste Lamarck, 1802). See also the English translation: *Hydrogeology* by J. B. Lamarck, trans. Albert V. Carozzi (Urbana: University of Illinois Press, 1964).

33. Jean-André Deluc, "Lettres à Delamétherie," 8th letter, "Sur quelques points fondamentaux relatifs à l'histoire ancienne de la terre," *Observations sur la Physique . . .*, 37 (1790): 203.

## Chapter 10: The Use of Fossils

1. Jean-André Deluc, *Lettres physiques et morales sur les montagnes et sur l'histoire de la terre et de l'homme, adressées à la reine de la Grande-Bretagne* (La Haye: chez de Tune, 1778), vii–viii; and his *Lettres physiques et morales sur l'histoire de la terre et de l'homme, adressées à la reine de la Grande Bretagne*, 5 vols. in 6 (La Haye: chez de Tune, 1779).

2. Jean-André Deluc, *Traité élémentaire de géologie* (Paris: Courcier, 1809).

3. Jean-André Deluc, "Lettres à Delamétherie," *Observations sur la physique . . .*, vols. 35–42 (1790–1793); see 18th letter, "Sur les Agathes, les Couches calcaires et sur une classe de couches d'argile," 39 (1791): 453–464.

4. Ibid., 24th letter, "Sur la nature du Silex et sur l'origine des substances minérales des Couches coquillières," 41 (1791): 45–46.

5. J.L.G. Soulavie, *Histoire naturelle de la France méridionale . . .*, 7 vols. (Nîmes: C. Belle, 1780–1784), 1:319.

6. G. Stegagno, *Il Veronese Giovanni Arduino e il suo contributo al progresso della scienza geologica* (Verona: Tipographia operaia, 1929), 14.

7. Jean-André Deluc, *Lettres sur l'histoire physique de la terre, adressées à M. le professeur Blumenbach, renfermant de nouvelles preuves géologiques et historiques de la mission divine de Moyse* (Paris: chez Nyon, 1798), 392.

8. Jean-André Deluc, "Lettres à Delamétherie," 12th letter, "Sur les Couches calcaires de la seconde classe . . ." 38 (1791): 100–101.

9. Soulavie, *Histoire naturelle*, 1:348.

10. Leopold von Buch, *Gesammelte Schriften*, 5 vols., ed. J. Ewald, J. Roth, H. Eck, and W. Dames (Berlin: G. Reimer, 1767–1885), 1:129.

11. G. P. Deshayes, "Tableau comparatif des espèces de coquilles vivantes avec les espèces de coquilles fossiles des terrains tertiaires de l'Europe et des espèces fossiles de ces terrains entre eux," *Bull. Société géologique de France* 1 (1831): 185.

12. Charles Lyell, *Principles of Geology; Being an Attempt to explain the Former Changes of the Earth's Surface, by Reference to Causes Now in Operation*, 3 vols. (London: J. Murray, 1830–1833), vol. 3.

13. Stegagno, *Il Veronese Giovanni Arduino*, 13. Vallisnieri's father, an important Italian naturalist, wrote *De' corpi marini che su'monti si trovano; della loro origine; e dello stato del mondo avanti'l Diluvio, nel Diluvio, e dopo il Diluvio* (On marine bodies found on mountains;

on their origin and the state of the world before the Flood, during the Flood, and after the Flood) (Venice: Domenico Loviso, 1721).

14. Georges Cuvier and Alexandre Brongniart, *Essai sur la géographie minéralogique des environs de Paris avec une carte géognostique, et des coupes de terrains* (Paris: Potey, 1811), 18.

15. Alexandre Brongniart, "Sur les caractères zoologiques des formations: Avec application de ces caractères à la détermination de quelques terrains de craie," *Annales des Mines* 6 (1821): 541–552.

16. Deshayes, quoted in Ami Boué, *Mémoires géologiques et paléontologiques*, 2 vols. (Paris: Ami Boué, 1832), 100.

17. Georges Cuvier, *Discours sur les révolutions de la surface du globe et sur les changements qu'elles ont produits dans le règne animal* (Paris: Dufour et d'Ocagne, 1825).

18. Alexandre Brongniart, "Sur les terrains qui paroissent avoir été formés sous l'eau douce," *Annales Muséum National Histoire Naturelle* 15 (1810): 357–405.

19. Cuvier, *Discours*, 129–130.

20. Ibid., 298–314.

21. William Smith, *Strata Identified by Organized Fossils, Containing Prints on Colored Paper of the Most Characteristic Specimens in Each Stratum* (London: W. Ardin, 1816); and his *Stratigraphical System of Organized Fossils, with Reference to the Specimens of the Original Geological Collection in the British Museum: Explaining Their State of Preservation and Their Use in Identifying the British Strata* (London: E. Williams, 1817).

22. John Phillips, *Memoirs of William Smith, LL.D., Author of the Map of the Strata of England and Wales* (London: J. Murray, 1844), 14. See also J.G.C.M. Fuller, "The Industrial Basis of Stratigraphy: John Strachey and William Smith," *Bulletin of the American Association of Petroleum Geologists* 53 (1969): 2256–2273.

23. Antoine Laurent Lavoisier, "Observations générales sur les couches modernes horizontales, qui ont été déposées par la mer, et sur les conséquences qu'on peut tirer de leurs dispositions, relativement à l'ancienneté du globe terrestre," *Mémoires de l'Académie Royale des Sciences, Paris* (written for the 1789 meeting, but published in 1793): 351–371. See also edited translation by A. V. Carozzi: "Lavoisier's Fundamental Contribution to Stratigraphy: General Observations on the Recent Marine Horizontal Beds and on Their Significance for the History of the Earth," *Ohio Journal of Science* 65 (1971): 71–85.

24. Rhoda Rappaport, "Lavoisier's Theory of the Earth," *British Journal for the History of Science* 6 (1973): 247–260.

25. Jean-Etienne Guettard, "Mémoire et carte minéralogique sur la nature et la situation des terreins qui traversent la France et l'Angleterre," *Mémoires de l'Académie Royale des Sciences, Paris* (1746): 363–392.

26. Bernard le Bouvier de Fontenelle, "Sur les Coquilles fossiles de Touraine," *Histoire de l'Académie Royale des Sciences, Paris* (1720): 11–12.

27. William Smith, *A Delineation of the Strata of England and Wales, with Part of Scotland; Exhibiting the Collieries and Mines, the Marshes and Fen Lands Originally Overflowed by the Sea . . . including 15 colored double maps* ([London] 1815).

## Chapter 11: Uniformitarianism versus Catastrophism

1. Charles Darwin, The Autobiography of Charles Darwin, 1809–1882, with original omissions restored, edited with appendix and notes by his granddaughter Nora Barlow (New York: Harcourt, Brace and Co., 1958), 69, 102–103.
2. Charles Lyell, Principles of Geology, Being an Attempt to Explain the Former Changes of the Earth's Surface, by Reference to Causes Now in Operation, 3 vols. (London: J. Murray, 1830–1833), vol. 1.
3. Ibid., vol. 3, 1833 edition, reprinted in The Sources of Science no. 84 (New York and London: Johnson Reprint Corp., 1969), 3:230. See also Leonard G. Wilson, Charles Lyell, the years to 1841: The Revolution in Geology (New Haven: Yale University Press, 1972), 203.
4. Charles Lyell, Elements of Geology (London: J. Murray, 1838).
5. Constant Prévost, "Sur un nouvel exemple de la réunion de coquilles marines et de coquilles fluviatiles dans les mêmes couches," Journal de Physique 92 (1821): 418–428.
6. Constant Prévost, "Observations sur les grès coquilliers de Beau-Champ et sur les mélanges de coquilles marines et fluviatiles dans les mêmes couches," Journal de Physique 94 (1822): 1–18.
7. Constant Prévost, "Les continents actuels ont-ils été à plusieurs reprises submergés par la mer?" Mémoire, Société Histoire Naturelle, Paris 4 (1828): 249, 346.
8. Constant Prévost, "Essai sur la formation des terrains des environs de Paris," lu à l'Académie des Sciences, juillet 1827, in Documents pour l'histoire des terrains tertiaires (n.p., n.d.), 93–124.
9. William Whewell, review of Principles of Geology, by Charles Lyell, Quarterly Review 47 (1832): 103–132.
10. Constant Prévost, "Réponse à Dufrénoy," Bull. Société géologique de France 3 (1833): 413.
11. Charles Darwin, "Observations of Proofs of Recent Elevation on the Coast of Chile," Proceedings, Geological Society of London 2 (1837): 446–449.
12. Martin J. S. Rudwick, "A Critique of Uniformitarian Geology: A Letter from W. D. Conybeare to Ch. Lyell, 1841," Proceedings, American Philosophical Society 111 (1967): 272–287.
13. Lyell, Principles of Geology, 2:179.
14. Constant Prévost, "Fossile," in Dictionnaire Universel d'Histoire naturelle, Paris, ed. Charles d'Orbigny (Paris: Renard, Martinet et Cie, 1841–1849), 5:684.
15. Constant Prévost, "De la chronologie des terrains et du synchronisme des formations," Comptes Rendus de l'Académie des Sciences, Paris 20 (1845): 1070.
16. Georges Cuvier, Discours sur les révolutions de la surface du globe et sur les changements qu'elles ont produits dans le règne animal (Paris: Dufour et d'Ocagne, 1825), 4.
17. Étienne Geoffroy Saint-Hilaire, Études progressives d'un naturaliste pendant les années 1834 et 1835, faisant suite à ses publications dans les 42 volumes des mémoires et annales du Muséum national d'histoire naturelle (Paris: chez Roret, Denain et Delamarre, 1835).
18. Jean-André Deluc, Lettres sur l'histoire physique de la terre, adressées

à M. le professeur Blumenbach, renfermant de nouvelles preuves géologiques et historiques de la mission divine de Moyse (Paris: chez Nyon, 1798), 60–61.

19. Cuvier, Discours, 27–28.

20. Charles Lyell, letter to Roderick Murchison, 17 January 1829, in Charles Lyell, the Years to 1841: The Revolution in Geology, by L. G. Wilson (New Haven: Yale University Press, 1972), 256.

21. Henri Ducrotay de Blainville, Ostéographie. . . . (Paris: Baillière, 1839–1864). See also P. Flourens, "Éloge historique de Marie Henri Ducrotay de Blainville" (Paris: Firmin-Didot frères, 1854), and extract in Mémoires de l'Académie des Sciences, Paris 27 (1860): i–xi.

22. Lyell, Principles of Geology, 10th ed. (London: J. Murray, 1867–1868), 1:133–134, 2:234–237.

23. Leopold von Buch, Travels through Norway and Lapland, during the years 1806, 1807 and 1808, trans. John Black (London: H. Colburn, 1813), chap. 9, 386–387.

24. Wilson, Charles Lyell, 385, 398–407.

25. L. Élie de Beaumont, "Rapport sur un mémoire de M. Bravais relatif aux lignes d'ancien niveau de la mer dans le Finmark," Comptes Rendus de l'Académie des Sciences, Paris 15 (1842): 807–849. Bravais was famous for his work in botany, physics, astronomy, and crystallography.

26. Lyell, Principles of Geology, 10th edition, 2:236.

27. Roderick Impey Murchison, "On the Geological Structure of the Alps, Apennines and Carpathians . . . ," Quarterly Journal of the Geological Society, London, 50 (1849): 157–312.

28. Charles Lyell, Elements of Geology, 6th edition (London: J. Murray, 1865), 50–52.

29. Lyell, Principles of Geology, 10th edition, 1:142–143.

30. T. Virlet d'Aoust, "Observations sur le métamorphisme normal et la probabilité de la non-existence de véritables roches primitives à la surface du globe," Bull. Société géologique de France, 2d ser., vol. 4 (1842): 505.

## Chapter 12: Catastrophes That Built the World

1. L. Élie de Beaumont, "Recherches sur quelques unes des révolutions de la surface du globe," in Henry Thomas De la Beche, Manuel géologique, trans. A. Brochant de Villiers (Paris: F. Levrant, 1833), 662. Hereafter cited as "Recherches," in Manuel géologique, by De la Beche.

2. Leopold von Buch, Observations sur les volcans d'Auvergne, trans. Mrs. de Kleinschord with notes by H. Lecoq (Clermont: Imprimerie de Thibaut-Landriot, 1842), 7.

3. Leopold von Buch, Description physique des Iles Canaries, suivie d'une indication des principaux volcans du globe, trans. C. Boulanger (Paris: Levrault, 1836), 167ff, 322ff.

4. Leopold von Buch, Gesammelte Schriften, 5 vols., ed. J. Ewald et al. (Berlin: G. Reimer, 1767–1885), 3:122.

5. Ibid., 218–221.

6. L. Élie de Beaumont, "Recherches sur quelques unes des révolutions de la surface du globe," Annales des Sciences Naturelles 18 (1829): 5–25, 284–416; 19 (1830): 5–99, 177–240.

7. L. Élie de Beaumont, "Recherches," in Manuel géologique, by De la

Beche, 617. See also the English translation of De la Beche's work: *A Geological Manual* (Philadelphia: Carey & Lea, 1832).

8. L. Élie de Beaumont, "Note sur les systèmes de montagnes les plus anciens d'Europe," *Bull. Société géologique de France*, 2d ser., vol. 4 (1847): 869–991.

9. L. Élie de Beaumont, *Notice sur les systèmes de montagnes*, 3 vols. (Paris: P. Bertrand, 1852).

10. L. Élie de Beaumont, "Extrait d'une lettre adressée à Prévost . . . ." *Comptes Rendus de l'Académie des Sciences, Paris* 21 (1850): 501–505.

11. Ami Boué, "Note sur les idées de M. de Beaumont relativement au soulèvement successif des diverses chaînes du globe," *Journal de Géologie* 3 (1831): 338–358.

12. Joseph Fourier, "Remarques générales sur les températures du globe terrestre et des espaces planétaires," *Annales de chimie et de physique* 27 (1824): 136–167.

13. Louis Cordier, "Essai sur la température de la terre," *Mémoire Muséum Histoire Naturelle* 15 (1827): 161–244.

14. L. Élie de Beaumont, "Recherches," in *Manuel géologique*, by De la Beche, 665.

15. Ibid., 618.

16. Ibid., 618–620.

17. Alexander von Humboldt, "Esquisse d'un tableau géologique de l'Amérique méridionale," *Journal de Physique* 53 (1801): 47.

18. Alexander von Humboldt, *Essai géognostique sur le gisement des roches dans les deux hémisphères* (Paris: F. G. Levrault, 1823), 60.

## Chapter 13: The Help of Biostratigraphy

1. For further details on stratigraphy see Marshall Kay and Edwin H. Colbert, *Stratigraphy and Life History* (New York: Wiley, 1965).

2. Alcide Dessaline d'Orbigny, *Cours élémentaire de paléontologie et de géologie stratigraphiques*, 3 vols. (Paris: Victor Masson, 1849–1852), 2:252.

3. Ibid., 1:135.

4. Ibid., 156.

5. Marcel de Serres, *De la cosmogonie de Moïse comparée aux faits paléontologiques* (Paris: Lagny Frères, 1838), 83.

6. P. Fischer, "Notice sur la vie et les travaux d'Alcide d'Orbigny," *Bull. Société géologique de France*, 3d ser., vol. 6 (1878): 434–453.

7. Amanz Gressly, "Geognostische Bemerkungen über den Jura der nordwestlichen Schweiz, besonders des Kantons Solothurn und der Grenz-Partien der Kantone Bern, Aargau und Basel," *Neues Jahrbuch für Mineralogie, Geognosie, und Petrefactenkunde* (1836), 659–675; and his "Observations géologiques sur le Jura soleurois," *Nouveaux Mémoires de la Société helvétique des Sciences naturelles, Neuchâtel* 2 (1838): 1–112. See Eugène Wegmann, "L'exposé original de la notion de faciès par A. Gressly (1814–1865)," *Sciences de la terre, Nancy* 9 (1962–1963): 83–119, with facsimile reproduction of parts of Gressly's definition of facies (cited by Wegmann on p. 11).

8. Eugène Renevier, "Tableau des terrains sédimentaires qui représentent les époques de la phase organique," *Bull. Société Vaudoise des Sciences naturelles, Lausanne* 13 (1874–1875): 218–252.

9. Eugène Renevier, "Chronographe géologique: Seconde édition du Tableau des terrains sédimentaires formés pendant les époques de la phase organique du globe terrestre, mis au point et entièrement retravaillé sur un plan nouveau avec texte explicatif suivi d'un répertoire stratigraphique polyglotte," *Congrès Géologique International, Comptes Rendus, sixième session, Zurich, 1894* (1897): 519–695. See also his "Résumé du chronographe géologique," *Eclogae Geol. Helvetiae* 5 (1897–1898): 69–75.

## Chapter 14: Unraveling the Earth's Age

1. Alexandre Brongniart, *Tableau des terrains qui composent l'écorce du globe: ou, Essai sur la structure de la partie connue de la terre* (Paris: F. G. Levrault, 1829).

2. Martin J. S. Rudwick, *The Great Devonian Controversy: The Shaping of Scientific Knowledge among Gentlemanly Specialists* (Chicago: University of Chicago Press, 1985).

3. William Buckland, *Vindiciae geologicae; or, The connexion of geology with religion, explained in an inaugural lecture delivered before the University of Oxford, May 15, 1819, on the endowment of a readership in geology by His Royal Highness the Prince Regent* (Oxford: Oxford University Press, 1820).

4. Rudwick, *Great Devonian Controversy*, 72–73.

5. Roderick Impey Murchison, *The Silurian System, founded on geological researches in the counties of Salop, Hereford, Radnor, Montgomery, Caermarthen, Brecon, Pembrocke, Monmouth, Gloucester, Worcester, and Stafford; with descriptions of the coalfields and overlying formations* (London: J. Murray, 1839).

6. Adam Sedgwick, "On the Silurian and Cambrian systems, exhibiting the order in which the older sedimentary strata succeed each other in England and Wales," *Transactions of the British Association for the Advancement of Science* (1835): 59–61.

7. Adam Sedgwick and Roderick Impey Murchison, "On the physical structure of Devonshire, and on the subdivisions and geological relations of its older stratified deposits," *Transactions of the Geological Society of London*, 2d ser., vol. 5 (1842): 633–704.

8. B. Balan, *L'Ordre et le Temps. L'anatomie comparée et l'histoire des vivants au XIXe siècle* (Paris: Brin, 1979), 389.

9. Joachim Barrande, *Système silurien du centre de la Bohème. Partie I: Recherches paléontologiques*, 8 vols. (Prague and Paris: Joachim Barrande, 1852), see vol. 1, *Texte, Crustacés: Trilobites*, Avant-propos xxi–xxx, and Introduction historique, 1–56.

10. John William Dawson, "On the Structure of Certain Organic Remains (*Eozoön canadense*) in the Laurentian Rocks of Canada," *Quarterly Journal of the Geological Society of London* 21 (1865): 51–59.

11. Émile Haug, *Les périodes géologiques*, vol. 2 of *Traité de géologie* (Paris: Armand Colin, 1908–1911), 569.

12. Charles Barrois, "Sur la présence de fossiles dans le terrain azoïque de Bretagne," *Comptes Rendus de l'Académie des Sciences, Paris* 115 (1892): 326–328. See also Lucien Cayeux, "Les preuves de l'existence d'organismes dans le terrain précambrien—Première note sur les Radio-

laires précambriens," *Bulletin, Société géologique de France*, 3d ser., vol. 22 (1894): 197–228.

13. François Ellenberger, "A l'aube de la géologie moderne: Henri Gautier (1660–1737)," *Histoire et Nature* 9–10 (1976–1977): 70.

14. Pierre Bernard Palassou, *Essai sur la minéralogie des Mont-Pyrénées*, 2d ed. (Paris: Didot jeune), xi.

15. Marcel de Serres, *De la cosmogonie de Moïse comparée aux faits paléontologiques* (Paris: Lagny Frères, 1838), 39.

16. William Buckland, *Geology and Mineralogy Considered with Reference to Natural Theology* (London: William Pickering, 1836). See French translation: *La géologie et la minéralogie dans leurs rapports avec la théologie naturelle*, trans. M. L. Doyère (Paris: Crochard, 1838), 18.

17. Martin J. S. Rudwick, "A Critique of Uniformitarian Geology: A Letter from W. D. Conybeare to Ch. Lyell, 1841," *Proceedings, American Philosophical Society* 111 (1967): 281.

18. Lyell, *Principles of Geology*, 10th edition, 1:300–301.

19. Lord Kelvin, "The Age of the Earth as an Abode Fitted for Life," *Annual Report of the Smithsonian Institution* (1897): 337–351. See also J. D. Burchfield, *Lord Kelvin and the Age of the Earth* (New York: Science History Publication, 1975).

20. John Joly, *Radioactivity in Geology: An Account of the Influence of Radioactive Energy on Terrestrial History* (London: A. Constable, 1909).

21. René Taton, ed., *Histoire générale des sciences*, 4 vols. in 3 (Paris: Presses universitaires de France, 1957–1964), 2:491.

## Chapter 15: The Breaking Up of the Crust

1. Eduard Suess, *Die Entstehung der Alpen* (Vienna: W. Braumüller, 1875).

2. Eduard Suess, *Das Antlitz der Erde*, 3 vols. (Prague: Tempsky, 1883–1909).

3. Marcel Bertrand, "Rapports de structure des Alpes de Glaris et du bassin houiller du Nord," *Bull. Société géologique de France*, 3d ser., vol. 12 (1883–1884): 318–330.

4. Marcel Bertrand, "La chaîne des Alpes et la formation du continent européen," *Bull. Société géologique de France*, 3d ser., vol. 15 (1887): 423–447.

5. Ibid., 432.

6. Ibid., 434, 438.

7. Eduard Suess, *The Face of the Earth*, 3 vols., trans. Hertha B. C. Sollas under the direction of W. J. Sollas (Oxford: Clarendon, 1906), vol. 2, pt. 3, chap. 14, 535–556; see in particular 553.

8. Charles Depéret and E. Cazist, "Note sur les gisements pliocènes et quaternaires marins des environs de Nice," *Bulletin, Société géologique de France*, 4th ser., vol. 3 (1903): 321–347.

9. Marcel Bertrand, *Mémoire sur les refoulements qui ont plissé l'écorce terrestre et sur le rôle des déplacements horizontaux* (Paris: Gauthier-Villars, 1908).

10. Pierre Termier, *A la gloire de la terre* (Paris: Desclée de Brouwer, 1922).

11. Émile Haug, "Les géosynclinaux et les aires continentales: Contri-

bution à l'étude des transgressions et régressions marines," *Bulletin, Société géologique de France*, 3d ser., 28 (1900): 617–711.

12. James Dana, "On Some Results of the Earth's Contraction from Cooling, Including a Discussion of the Origin of Mountains and the Nature of the Earth's Interior," part 1, *American Journal of Science* 5 (1873): 423–443; and his *Manual of Geology Treating of the Principles of the Science, with Special Reference to American Geological History, for the Use of Colleges, Academia, and Schools of Science* (Philadelphia: T. Bliss, 1862).

13. James Hall, *Paleontology of New York*, 8 vols. (Albany, 1859), 3, pt. 1.

14. Émile Haug, *Traité de géologie*, 2 vols. (Paris: Armand Colin, 1907–1911).

15. Henry Clifton Sorby, *Sorby on Sedimentology: A Collection of Papers from 1851–1908, in Geological Milestones*, vol. 1, ed. Charles H. Summerson (Miami: Comparative Sedimentology Laboratory, Division of Marine Geology and Geophysics, University of Miami, Rosenstiel School of Marine & Atmospheric Science, 1976). See also *Sorby on Geology: A Collection of Papers from 1853–1906, in Geological Milestones*, vol. 3, ed. Charles H. Summerson (Miami: idem, 1978).

16. Ferdinand Fouqué and Auguste Michel Lévy, *Introduction à l'étude des roches éruptives françaises: Minéralogie micrographique, Mémoire pour servir à l'explication de la carte géologique détaillée de la France* (Paris: Imprimerie A. Quantin, 1978), 156–157.

## Chapter 16: Continental Drift

1. The printed version of Wegener's paper is entitled "Die Entstehung der Kontinente," *Petermanns Mitteilungen* (1912): 185–195, 253–256, 305–309.

2. Alfred Wegener, *Die Entstehung der Kontinente und Ozeane* (Braunschweig: Sammlung Vieweg, no. 23, 1915).

3. Martin Schwarzbach, *Alfred Wegener und die Drift der Kontinente* (Stuttgart: Wissenschaftliche Verlagsgesellschaft, 1980), 42.

4. Ibid., 46–47.

5. Albert V. Carozzi, "The Reaction in Continental Europe to Wegener's Theory of Continental Drift," *Earth Sciences History* 4 (1985): 122–137. See also Naomi Oreskes, "The Rejection of Continental Drift," *Historical Studies in the Physical Sciences* 18 (1988): 311–348; and Ursula B. Marvin, "The British Reception of Alfred Wegener's Continental Drift Hypothesis," *Earth Sciences History* 4 (1985): 138–159.

6. Alfred Wegener, *The Origin of Continents and Oceans*, trans. John Biram from 4th rev. German edition (New York: Dover, 1966), 1.

7. Antonio Snider-Pellegrini, *La Création et ses Mystères dévoilés, Ouvrage où l'on expose l'origine de l'Amérique et de ses habitants primitifs* (Paris: Franck et Dentu, 1858). See also Albert V. Carozzi, "A propos de l'origine de la théorie des dérives continentales: Francis Bacon (1620), François Placet (1668), A. von Humboldt (1801) et A. Snider (1858)," *Comptes Rendus, Société de Physique et d'Histoire Naturelle de Genève* 4 (1969): 171–179.

8. Albert V. Carozzi, "New Historical Data on the Origin of the Theory of Continental Drift," *Bulletin of the Geological Society of America* 81 (1970): 283–286.

9. Wegener, *Origin of Continents and Oceans*, 102.
10. Ibid., 17.
11. Ibid., 102–103.
12. R. Staub, "Der Bau der Alpen," *Beiträge zur geologischen Karte der Schweiz*, N. F. (Bern: A. Francke S.A., 1924).
13. Wegener, *Origin of Continents and Oceans*, 10–11.
14. Ibid., 12.
15. John Henry Pratt, "On the Attraction of the Himalaya Mountains and the Elevated Regions beyond Them, Upon the Plumb-Line in India," *Philosophical Transactions, Royal Society of London*, 145 (1855): 53–100.
16. George Biddel Airy, "On the Computation of the Effect of Attraction of Mountain-Masses, as Disturbing the Apparent Astronomical Latitude of Stations in Geodetic Surveys," *Philosophical Transactions, Royal Society of London* 145 (1855): 101–104.
17. John Henry Pratt, "On the Deflection of the Plumb-Line in India, Caused by the Attraction of the Himalaya Mountains and of the Elevated Regions Beyond; and Its Modification by the Compensating Effect of a Deficiency of Matter below the Mountain Mass," *Philosophical Transactions, Royal Society of London* 49 (1859): 745–778; and his "On the Influence of the Ocean on the Plumb-Line in India," *Philosophical Transactions, Royal Society of London* 49 (1859): 779–796. See also Mott T. Greene, *Geology in the Nineteenth Century: Changing Views of a Changing World* (Ithaca: Cornell University Press, 1982), 238–243.
18. C. E. Dutton, "On some Greater Problems of Physical Geology," *Bulletin of the Philosophical Society of Washington* 11 (1889): 51–64.
19. Wegener, *Origin of Continents and Oceans*, 43.
20. Bailey Willis, "Principles of Paleogeography," *Science* n.s. 31 (1910): 243.
21. J. D. Dana, "On the Volcanoes of the Moon," *American Journal of Science* 2 (1846): 335–355.
22. Ursula B. Marvin, *Continental Drift: The Evolution of a Concept* (Washington, D.C.: Smithsonian Institution Press, 1973), 38–39.
23. H. E. Le Grand, *Drifting Continents and Shifting Theories* (New York: Cambridge University Press, 1988), passim.
24. Wegener, *Origin of Continents and Oceans*, 21.
25. Ibid., 167.
26. Léonce Joleaud, "Essai sur l'évolution des milieux géophysiques et biogéographiques (A propos de la théorie de Wegener sur l'origine des continents)," *Bull. Société géologique de France*, 4th ser., vol. 23 (1923): 205–270.
27. P. Lake, "Wegener's Hypothesis of Continental Drift," *Geographical Journal* 61 (1923): 179–194.
28. W.A.J.M. van Waterschoot van der Gracht, ed., *Theory of Continental Drift: A Symposium* (Tulsa: American Association of Petroleum Geologists, 1928), 35–52.
29. Hans Stille, "Kordillerisch-atlantische Wechselbeziehungen," *Geologische Rundschau* 30 (1939): 315–342.
30. E. B. Taylor, "Bearing of the Tertiary Mountain Belts on the Origin of the Earth's Plan," *Bulletin of the Geological Society of America*, 21 (1910): 179–226.

31. Émile Argand, "La tectonique de l'Asie," *Comptes Rendus 13e Congrès Géologique International, Belgique* (1922): 171–372. See also translation by Albert V. Carozzi, *Tectonics of Asia* (New York: Hafner, 1977).
32. Alexander L. Du Toit, *Our Wandering Continents: An Hypothesis of Continental Drifting* (Edinburgh: Oliver and Boyd, 1937).
33. Arthur Holmes, "Radioactivity and Earth Movements," *Transactions, Geological Society of Glasgow* 18 (1930): 559–606.
34. Arthur Holmes, *Principles of Physical Geology*, 2d ed. (New York: Ronald Press, 1964), 1204.
35. R. Schwinner, "Vulkanismus und Gebirgsbildung," *Zeitschrift für Vulkanologie* 5 (1920): 176–230.

## Chapter 17: The Birth of the Oceans

1. Robert M. Wood, *The Dark Side of the Earth* (London: Allen and Unwin, 1985), 125–132. Wood cites, in particular, E. C. Bullard, "The Emergence of Plate Tectonics: A Personal View," *Annual Review of Earth and Planetary Sciences* 3 (1975): 1–30.
2. Jean-Pierre Rothé, "La zone séismique médiane indo-atlantique," *Proceedings, Royal Society of London* 222 (1954): 387–397.
3. Xavier Le Pichon, "La naissance de la tectonique des plaques," *La Recherche* 153 (1984): 416.
4. Ibid., 417.
5. Harry H. Hess, "The History of Ocean Basins," in *Petrologic Studies: A Volume to Honor A. F. Buddington,* ed. A.E.J. Engel, H. L. James, and B. F. Leonards (Boulder, Colo.: Geological Society of America, 1962), 599–620.
6. R. S. Dietz, "Continent and Ocean Basin Evolution by Spreading of the Sea Floor," *Nature* 190 (1961): 30–41.
7. A. Mohorovičić, "Das Beben vom 8.X.1909," *Jahrbuch Meteorologischen Observationen, Zagreb (Agram)* 9 (1909): 1–63.
8. H. Frankel, "The Development, Reception, and Acceptance of the Vine-Matthews-Morley Hypothesis," *Historical Studies in the Physical Sciences* 13 (1982): 1–39.
9. F. J. Vine and D. H. Matthews, "Magnetic Anomalies over Oceanic Ridges," *Nature* 199 (1963): 947–949.
10. K. Wadati, "On the Activity of Deep-focus Earthquakes in the Japan Islands and Neighbourhoods," *Geophysical Magazine* 8 (1935): 305–325.
11. H. Benioff, "Orogenesis and Deep Crustal Structure: Additional Evidence from Seismology," *Bulletin of the Geological Society of America* 65 (1954): 385–400.
12. David A. White, Dietrich H. Roeder, Thomas H. Nelson, and John C. Crowell, "Subduction," *Bulletin of the Geological Society of America* 81 (1970): 3431–3432.
13. Ursula B. Marvin, *Continental Drift: The Evolution of a Concept* (Washington: Smithsonian Institution Press, 1973), 136–137. See also A. Hallam, *A Revolution in the Earth Sciences: From Continental Drift to Plate Tectonics* (Oxford: Clarendon, 1973); and H. E. Le Grand, *Drifting Continents and Shifting Theories* (New York: Cambridge University Press, 1988).
14. Tuzo Wilson, "A New Class of Faults and Their Bearing on Continental Drift," *Nature* 207 (1965): 343–347.

15. W. J. Morgan, "Rises, Trenches, Great Faults, and Crustal Blocks," *Journal of Geophysical Research* 73 (1968): 1959–1982.
16. D. P. McKenzie and R. L. Parker, "The North Pacific: An Example of Tectonics on a Sphere," *Nature* 216 (1967): 1276–1280.
17. Xavier Le Pichon, "Sea-floor Spreading and Continental Drift," *Journal of Geophysical Research* 73 (1968): 3661–3697.
18. Niles Eldredge and Stephen Jay Gould, "Punctuated Equilibria: An Alternative to Phyletic Gradualism," in *Models in Paleobiology*, ed. J. M. Schopf, 82–115 (San Francisco: Freeman, Cooper & Co., 1972).
19. G. Einsele and A. Seilacher, eds., *Cyclic and Event Stratification* (Berlin: Springer-Verlag, 1982). See also Kathleen M. Marsaglia and George deVries Klein, "The Paleogeography of Paleozoic and Mesozoic Storm Depositional Systems," *Journal of Geology* 91 (1983): 117–142.
20. Hans Stille, "Kordillerisch-atlantische Wechselbeziehungen," *Geologische Rundschau* 30 (1939): 315–342.
21. Claude Allègre, *The Behavior of the Earth: Continental and Seafloor Mobility*, trans. D. Kurmes Van Dam (Cambridge: Harvard University Press, 1988).
22. C. R. Scotese, R. K. Bambach, C. Barton, R. Van der Voo, and A. M. Ziegler, "Paleozoic Base Maps," *Journal of Geology* 87 (1979): 217–227.
23. Tuzo Wilson, *Adventures in Earth History* (San Francisco: W. H. Freeman, 1970), 351.
24. Luis W. Alvarez, Walter Alvarez, Frank Asaro, and Helen V. Michel, "Extraterrestrial Cause for Cretaceous-Tertiary Extinction," *Science* 208 (1980): 1095–1108.
25. L. T. Silver and P. H. Schultz, eds., *Geological Implications of Impacts of Large Asteroids and Comets on the Earth*, Geological Society of America, Special Paper no. 190 (1982).
26. D. M. Raup and J. J. Sepkoski, Jr., "Periodicity of Extinctions in the Geologic Past," *Proceedings of the Natural Academy of Sciences, Washington* 81 (1984): 801–805. See also J. J. Sepkoski, Jr., "Global Bioevents and the Question of Periodicity," in *Global Bio-Events: A Critical Approach*, ed., O. H. Walliser, 47–61, Lecture Notes in Earth Sciences no. 8 (Berlin: Springer-Verlag, 1986); Sepkoski's "Periodicity of Extinction: A 1988 Update," in *Global Catastrophes in Earth History: An Interdisciplinary Conference on Impacts, Volcanism, and Mass Mortality*, Lunar and Planetary Institute, contribution 673 (Houston, 1988), 170–171; and Claude C. Albritton, Jr., *Catastrophic Episodes in Earth History* (London: Chapman and Hall, 1989).

# Glossary

The technical terms defined in this glossary are taken mostly from the *Glossary of Geology*, ed. Robert L. Bates and Julia A. Jackson, 3d ed. (American Geological Institute, 1987), with permission. For additional information, the reader is referred to the same publication.

**Altimeter**. An instrument, usually an aneroid barometer, for determining height above ground or above mean sea level, based on the decrease in atmospheric pressure accompanying an increase in elevation.

**Ammonite**. The coiled, chambered fossil shell of a cephalopod, called at first "horn of Ammon" because of its resemblance to the horn of Jupiter Ammon.

**Angular unconformity**. An unconformity in which younger sediments rest upon the eroded surface of tilted or folded older rocks.

**Asthenosphere**. The layer or shell of the earth below the lithosphere, which is weak and in which isostatic adjustments take place, magmas may be generated, and seismic waves are strongly attenuated.

**Azoic**. The earliest part of Precambrian time, represented by rocks in which there is no trace of life.

**Basalt**. Dark-colored igneous rocks, forming lava flows or small intrusions, composed chiefly of calcic feldspars and pyroxenes (iron-rich minerals), often displaying columnar structures.

**Bathymetric studies (bathymetry).** The measurement of ocean depths and the charting of the topography of the ocean floor.

**Belemnite.** A conical fossil, several inches long, consisting of the internal calcareous rod of an extinct animal allied to the cuttlefish.

**Biocenosis.** A group of organisms that live closely together and form a natural ecologic unit.

**Biofacies.** A distinctive assemblage of organisms formed under one set of environmental conditions, as compared with another assemblage formed at the same time but under different conditions.

**Biostratigraphy.** The separation and differentiation of rock units on the basis of the description and study of the fossils they contain.

**Caldera.** A large, basin-shaped volcanic depression, more or less circular in form, the diameter of which is many times greater than that of the included vent or vents.

**Caledonian orogeny.** The early Paleozoic orogeny (near the end of the Silurian) in western Europe that created a mountain belt, extending from Ireland and Scotland northeastward through Scandinavia.

**Cambrian.** The earliest period of the Paleozoic era, between approximately 570 and 500 million years ago.

**Carboniferous.** The Mississippian and Pennsylvanian periods combined, ranging from about 345 to about 280 million years ago. In European usage, the Carboniferous is considered a single period.

**Cenozoic.** An era of geologic time beginning about 65 million years ago, following the Mesozoic era. It is characterized by the evolution and abundance of mammals, advanced mollusks, birds, and angiosperms.

**Chalk.** An Upper Cretaceous soft, friable, fine-textured limestone formed by shallow-water accumulations of calcareous tests of floating microorganisms (chiefly foraminifers and algae). The best known and most widespread chalks are exposed in cliffs on both sides of the English Channel.

**Convection currents.** A supposed mass movement of subcrustal or mantle material, either laterally or in upward- or downward-directed convection cells, mainly as a result of heat variations.

**Cretaceous.** The final period of the Mesozoic era, ranging between 135 and 65 million years ago. It is named after the Latin word for chalk (*creta*) because of the English chalk beds of this age.

**Detrital rocks.** Rocks consisting of small particles, worn or broken away from a mass as by the action of water or glacial ice.

**Devonian.** A period of the Paleozoic era (after the Silurian and before the Mississippian) thought to have covered the span of time between 395 and 345 million years ago.

**Diagenesis.** All the chemical, physical, and biologic changes undergone by a sediment after its initial deposition, and during and after its lithification, exclusive of surficial alteration (weathering) and metamorphism.

**Discontinuity, seismic.** A surface at which seismic-wave velocities abruptly change; a boundary between seismic layers of the earth.

**Dolostone.** Rock consisting of the mineral dolomite, a calcium magnesium carbonate.

**Eclogite.** A granular rock composed essentially of garnet and sodic pyroxene. Rutile, kyanite, and quartz are typically present.

**Eocene.** An epoch of the early Cenozoic era, after the Paleocene and before the Oligocene.

**Eurasia.** Europe and Asia considered together as one continent.

**Eustatic.** Pertaining to worldwide changes in sea level that affect all the oceans. Eustatic changes have various causes, among which the most important are changes of volume of oceanic basins as a result of variations of new oceanic crust production and behavior, and the melting or increase of continental ice caps.

**Evaporite.** A nonclastic sedimentary rock composed primarily of minerals produced from a saline solution as a result of extensive or total evaporation of the solvent. Examples include gypsum, anhydrite, salt, and potassic salts.

**Facies.** A distinctive rock type, broadly corresponding to a certain environment or mode of deposition; e.g., "red-bed facies," "black-shale facies."

**Facies fossil.** A fossil, usually a single species or a genus, that is restricted to a defined stratigraphic facies or is adapted to life in a restricted environment. It prefers certain ecologic surroundings and may exist in them with little change for long periods of time.

**Fixist.** A believer in the permanence of oceans in contrast to a scientist in favor of continental drift (mobilist).

**Foci (focus).** The initial rupture of an earthquake, where strain energy is first converted to elastic wave energy; the point within the earth that is the center of an earthquake.

**Foreland.** A stable area marginal to an orogenic belt, toward which the rocks of the belt were thrust or overfolded. Generally the foreland is a continental part of the crust and the edge of the craton (a relatively rigid and immobile region).

**Formation.** A body of rock identified by lithic characteristics and stratigraphic position; it is mappable in the field or traceable in the subsurface. A formation name normally consists of a geographic name followed by a descriptive geologic term (usually the dominant rock type), e.g., Dakota Sandstone, Morrison Formation.

**Geosyncline.** A mobile downwarping of the crust of the earth, either elongate or basinlike, measured in scores of kilometers, in which sedimentary and volcanic rocks accumulate to thicknesses of thousands of meters. A geosyncline may form in part of a tectonic cycle in which orogeny follows.

**Gondwana.** The Late Paleozoic continent of the Southern Hemisphere. It was named by Suess for the Gondwana system of India, which has an age range from Carboniferous to Jurassic and contains glacial tillite in its lower part and coal measures higher up. Similar sequences of the same age are found in all the continents of the hemisphere; this similarity indicates that all these continents were once joined into a single larger mass. The counterpart in the Northern Hemisphere was Laurasia.

**Granite.** A coarse-grained igneous rock composed chiefly of quartz, potassium feldspar, plagioclase, biotite, and amphibole.

**Grauwacke.** Also called "graywacke," an old miner's term from the Harz Mountains in Germany, now generally applied to a dark gray, coarse-grained sandstone that consists of poorly sorted angular grains of quartz and feldspar, with a variety of rock fragments embedded in a compact argillaceous matrix.

**Gravitational sliding (gravity thrust).** Downward movement of rock masses on slopes by the force of gravity, e.g., along a thrust-fault plane.

**Greenhouse effect.** The heating of the earth's surface because outgoing long-wavelength terrestrial radiation is absorbed and reemitted by the carbon dioxide and water vapor in the lower atmosphere and eventually returns to the surface.

**Guyot.** A flat-topped seamount found chiefly in the Pacific Ocean, named after Arnold Guyot (1807–1884), a Swiss-American geologist.

**Hercynian mountains.** A European mountain chain formed during the late Paleozoic era, extending through the Carboniferous and Permian. The Hercynian orogeny is synonymous to the Variscan orogeny.

**Hiatus.** A break or interruption in the continuity of the geological record, such as the absence in a stratigraphic sequence of rocks that would normally be present but either were never deposited or were eroded before deposition of the overlying beds.

**Hinterland.** An area bordering, or within, an orogenic belt on the internal side, away from the direction of overfolding and thrusting.

**Igneous rocks.** They constitute one of the three main classes into which rocks are divided, the others being metamorphic and sedimentary. Igneous rocks solidified from molten or partly molten material, that is, from magma.

**Index fossil.** A fossil that identifies and dates the strata or succession of strata in which it is found. It is generally a genus, rarely a species, that combines morphologic distinctiveness with relatively common occurrence or great abundance and that is characterized by a broad, even worldwide, geographic range and by a narrow or restricted stratigraphic range. The best index fossils include swimming and floating organisms that evolved rapidly and were distributed widely, such as graptolites and ammonites.

**Induration.** The hardening of a sediment by heat, pressure, or the introduction of cementing material, especially the process by which sediments are converted into compact rocks.

**Iridium (Ir).** An element of the platinum group, rare on earth and common in meteorites.

**Isostasy.** The condition of equilibrium, comparable to floating, of the units of the lithosphere above the asthenosphere.

**Jurassic.** The second period of the Mesozoic era (after the Triassic and before the Cretaceous), thought to have covered the span of time between 190 and 136 million years ago. Named after the Jura Mountains in which rocks of this age were first studied.

**Klippe.** An isolated relief that is an erosional remnant or outlier of a nappe or overthrust.

**Laramide phase (or orogeny).** A time of deformation, typically recorded in the eastern Rocky Mountains of the United States, whose several phases extended from Late Cretaceous until the end of the Paleocene.

**Laurasia.** The protocontinent of the Northern Hemisphere, corresponding to Gondwana in the Southern Hemisphere, from which the present continents of the Northern Hemisphere (North America, Greenland, most of Eurasia, including India) have been separated by continental drift.

**Lignite.** A low-grade brownish black coal that is intermediate between peat and subbituminous coal.

**Lithofacies.** A lateral, mappable subdivision of a designated stratigraphic unit, distinguished from adjacent subdivisions on the basis of lithology.

**Lithosphere.** The solid portion of the earth; a layer of strength relative to the underlying asthenosphere for deformation at geologic rates. It includes the crust and part of the upper mantle.

**Lithostratigraphy.** The branch of stratigraphy that deals with the lithology of strata and with their organization into units based on lithologic character.

**Littoral.** Pertaining to the ocean environment between high water and low water, also called intertidal.

**Long-range correlation.** Demonstration of the equivalence of two or more geologic phenomena in different, far-apart areas.

**Magmatism.** The development and movement of magma, and its solidification to igneous rock.

**Mesozoic.** An era of geologic time from the end of the Paleozoic to the beginning of the Cenozoic, or from about 225 to about 65 million years ago.

**Metamorphic rocks.** Any rocks derived from preexisting rocks by mineralogical, chemical, and/or structural changes, essentially in the solid state, in response to marked changes in temperature, pressure, shearing stress, and chemical environment, generally at depth in the earth's crust.

**Metamorphism.** The mineralogical, chemical, and structural adjustment of solid rocks to physical and chemical conditions that have generally been imposed at depth below the surface zones of weathering and cementation, and which differ from the conditions under which the rocks in question originated.

**Micaschists.** Schists whose essential constituents are mica and quartz, and whose schistosity is mainly due to the parallel arrangement of mica flakes.

**Mid-Atlantic Ridge.** A north-south suboceanic ridge in the Atlantic Ocean, from Iceland to Antarctica, on whose crest are several groups of islands. According to plate tectonics, the axis along which North America has split away from Eurasia, and South America from Africa and where new oceanic crust is upwelling.

**Miocene.** An epoch of the Cenozoic era, after the Oligocene and before the Pliocene.

**Nappe.** A sheetlike rock unit that has moved on a predominantly horizontal surface. The mechanism may be thrust faulting, recumbent folding, or both.

**Oceanic trench.** A narrow, elongate depression of the deep-sea floor, commonly a subduction zone, with steep sides, oriented parallel to the trend of the continent and between the continental margin and the abyssal hills.

**Oligocene.** An epoch of the Cenozoic era, after the Eocene and before the Miocene.

**Ordovician.** The second earliest period of the Paleozoic era (after the Cambrian and before the Silurian), thought to have covered the span of time between 500 and 430 million years ago.

**Orogenic belt.** A linear or arcuate region that has been subjected to folding and other deformation during an orogenic cycle. Orogenic belts are mobile belts during their formative stages, and most of them later become mountain belts.

**Orogeny.** Literally, the process of formation of mountains. The term came into use in the middle of the nineteenth century, when the process was thought to include both the deformation of rocks within the mountains and the creation of the mountainous topography. Only much later was it realized that the two processes were not closely related either in origin or in time. Today, most geologists regard the formation of mountainous topography as postorogenic. By present geological usage, orogeny is the process by which structures within fold-belt mountainous areas were formed, including thrusting, folding, and faulting in the outer and higher layers, and plastic folding, metamorphism, and plutonism in the inner and deeper layers. Only in the very youngest, late Cenozoic mountains is there any evident causal relation between rock structure and surface landscape. Little such evidence is available for the early Cenozoic, still less for the Mesozoic and Paleozoic, and virtually none for the Precambrian—yet all the deformational structures are much alike, whatever their age, and are appropriately considered as products of orogeny.

**Outcrop.** That part of a geological formation or structure that appears at the surface of the earth.

**Overthrust.** A low-angle thrust fault of large scale, with displacement generally measured in kilometers.

**Paleocene.** The first epoch of the Cenozoic era.

**Paleomagnetism.** The study of natural remanent magnetization in order to determine the intensity and direction of the earth's magnetic field in the geologic past.

**Paleozoic.** An era from the end of the Precambrian to the beginning of the Mesozoic, or from about 570 to about 225 million years ago.

**Pangea.** A supercontinent that existed from about 300 to about 200 million years ago and included most of the continental crust of the earth from which the present continents were derived by fragmentation and continental displacement.

**Pelagic.** Pertaining to the environment of the open ocean rather than the bottom or shore areas.

**Peneplain.** A low, nearly featureless, gently undulating land surface, which presumably has been produced by the process of subaerial erosion almost to base level, as well as by marine, eolian, and even glacial erosion.

**Permian.** The last period of the Paleozoic era (after the Pennsylvanian), thought to have covered the span of time between 280 and 225 million years ago.

**Phanerozoic.** That part of geologic time represented by rocks in which the evidence of life is abundant, i.e., Cambrian and later time.

**Phyllite.** A metamorphosed rock, intermediate in grade between slate and micaschist. Minute crystals of sericite and chlorite impart a silky sheen to the surface of cleavage (or schistosity).

**Plate tectonics.** A theory of global tectonics in which the lithosphere is divided into a number of plates whose pattern of horizontal movement is that of torsionally rigid bodies that interact with one another at their boundaries, causing seismic and tectonic activity along these boundaries.

**Pleistocene.** An epoch of the Cenozoic era, after the Pliocene and before the Recent. It began 2.5 million years ago and lasted until some 10,000 years ago.

**Pliocene.** An epoch of the Cenozoic era, after the Miocene and before the Pleistocene.

**Precambrian.** All geologic time and its corresponding rocks before the beginning of the Paleozoic; it is equivalent to about 90 percent of geologic time.

**Pyroxene porphyry.** An igneous rock with conspicuous crystals of pyroxene (a dark-colored ferro-magnesian silicate) scattered in a fine-grained groundmass.

**Quaternary.** Originally the name given to the geologic time of the Pleistocene and the Recent, which were considered periods instead of epochs. If still used in that sense, the Quaternary is the second period of the Cenozoic era, following the Tertiary.

**Recumbent fold.** An overturned fold, the axial surface of which is horizontal or nearly so.

**Regression.** The retreat of the sea from land areas, and the consequent evidence of such withdrawal (such as fall of sea level or uplift of land), that brings nearshore, typically shallow-water environments to areas formerly occupied by offshore, typically deep-water conditions, or that shifts the boundary between marine and nonmarine deposition (or between deposition and erosion) toward the center of a marine basin.

**Remanent magnetization.** That component of a rock's magnetization that has a fixed direction relative to the rock and is independent of moderate, applied magnetic fields such as the earth's magnetic field.

**Rift (World Rift System).** A major tectonic element of the earth, consisting of midoceanic ridges and their associated rift valleys, such as those along the Mid-Atlantic Ridge. It is believed to be the locus of the extensional splitting and upwelling of magma that has resulted in sea-floor spreading.

**Schist.** A strongly foliated crystalline rock, formed by dynamic metamorphism, that can be readily split into thin flakes or slabs due to the well-developed parallelism of more than 50% of the minerals present, particularly those of lamellar or elongate prismatic mica and hornblende.

**Sea-floor spreading.** A hypothesis that the oceanic crust is increasing by convective upwelling of magma along the midoceanic ridges or world rift system, and by a moving-away of the new material at a rate of one to ten centimeters per year. This movement provides the source of dynamic thrust in the hypothesis of plate tectonics.

**Sediment.** Solid fragmental material that originates from weathering of rocks and is transported or deposited by air, water, or ice, or that accumulates by other natural agents, such as chemical precipitation from solution or secretion by organisms, and that forms in layers on the earth's surface at ordinary temperatures in a loose, unconsolidated form; e.g., terrigenous gravels, sands, silts, and muds, as well as carbonate muds.

**Sedimentary rock.** A rock resulting from the consolidation of loose sediment that has accumulated in layers, e.g., a conglomerate consisting of mechanically formed fragments of older rock, transported from its source and deposited in water, or from air or ice; or a chemical rock such as rock salt or gypsum, formed by precipitation from solution; or an organic rock such as certain limestones, consisting of the remains or secretions of algae and animals.

**Sedimentation.** The process of deposition of sediment.

**Seismic belt.** An elongate earthquake zone, especially a zone of subduction or a midoceanic ridge where oceanic crust is being produced.

**Silurian.** A period of the Paleozoic, thought to have covered the span of time between 430 and 395 million years ago. The Silurian follows the Ordovician and precedes the Devonian.

**Stage.** A chronostratigraphic unit of smaller scope and rank than a series (a major unit of chronostratigraphic correlation). Most stage names are based on lithostratigraphic units, although preferably a stage should have a geographic name not previously used in stratigraphic nomenclature, such as Kimmeridgian or Oxfordian.

**Stromatolite.** A laminated organosedimentary structure, forming mats and mounds, produced by sediment trapping, binding, and/or precipitation as a result of the growth and metabolic activity of microorganisms, principally blue-green algae, ranging in age from Precambrian to Present.

**Subduction.** The process of one lithospheric plate descending beneath another.

**Subsidence.** A sinking or downwarping of a large part of the earth's crust relative to its surrounding parts, such as the formation of a rift valley or the lowering of a coast due to tectonic movements.

**Taconic phase (orogeny).** In the latter part of the Ordovician period, named for the Taconic Range of eastern New York State, well developed through most of the northern Appalachians in the United States and Canada.

**Tectonics.** A branch of geology dealing with the broad architecture of the outer part of the earth; that is, the regional assembling of structural or deformational features, a study of their mutual relations, origin, and historical evolution.

**Tempestite.** A storm deposit, showing evidence of violent disturbance of preexisting sediments followed by their rapid redeposition, all in a shallow-water environment.

**Tertiary.** Initially considered an era encompassing the geologic time from the Paleocene to the Pliocene, which were considered periods instead of epochs. If still used in that sense, the Tertiary is the first period of the Cenozoic era, followed by the Quaternary.

**Thrust.** An overriding movement of one crustal unit over another, as in thrust faulting.

**Transform fault.** A variety of a strike-slip fault along which the displacement suddenly stops or changes form. Many transform faults are associated with midoceanic ridges; they also correspond to a plate boundary that shows pure strike-slip displacement.

**Transgression.** The spread or extension of the sea over land areas, and the consequent evidence of such advance. Also, any change (such as rise of sea level or subsidence of land) that brings offshore, typically deep-water environments to areas formerly occupied by nearshore, typically shallow-nonmarine deposition.

**Triassic.** The first period of the Mesozoic era (after the Permian of the Paleozoic era and before the Jurassic), thought to have covered the span of time between 225 and 190 million years ago.

**Trilobite.** Any marine arthropod of the extinct class Trilobita, from the Paleozoic era, having a flattened, trilobate oval body varying in length from one inch or less to two feet.

**Tsunami.** A gravitational sea wave (erroneously called tidal wave) produced by any large-scale, short-duration disturbance of the ocean floor, principally by submarine earthquakes. It is characterized by great speed of propagation (up to 950 km/hour), long wavelength (up to 200 km), long period (generally 10–60 minutes), and low observable amplitude on the open sea.

**Unconformable.** Said of strata or stratification exhibiting the relation of unconformity to the older underlying rocks; not succeeding the underlying rocks in immediate order of age or not fitting together with them as parts of a continuous whole.

**Uniformitarianism.** The fundamental principle that geologic processes and natural laws now operating to modify the earth's crust have acted in the same regular manner and with essentially the same intensity throughout geologic time, and that past geologic events can be explained by phenomena and forces observable today according to the classical concept that the present is the key to the past.

**Vitrification.** Process which melts mineral substances into a glassy mass.

# Index of Names

Lightning Source UK Ltd.
Milton Keynes UK
UKOW02f1959250516

275001UK00001B/131/P